Ben Mezrich

THE accidental billionaires

Ben Mezrich, a Harvard graduate, is the author of eleven books, including the international bestseller *Bringing Down the House*, which spent sixty-three weeks on the *New York Times* bestseller list and was made into the movie *21*, starring Kevin Spacey. He is a columnist for *Boston Common* and a contributor to *Flush* magazine. Ben lives in Boston with his wife, Tonya.

www.benmezrich.com

THE
accidental billionaires

ALSO BY BEN MEZRICH

Rigged

Busting Vegas

Ugly Americans

Bringing Down the House

The Carrier

Skeptic

Fertile Ground

Skin [an *X-Files* book]

Reaper

Threshold

THE accidental billionaires

▶ THE FOUNDING OF FACEBOOK

A TALE OF SEX | MONEY | GENIUS | AND BETRAYAL

Ben Mezrich

ANCHOR BOOKS | A DIVISION OF RANDOM HOUSE, INC. | NEW YORK

FIRST ANCHOR BOOKS EDITION, MAY 2010

 Published in the United States by Anchor Books, a division of Random House, Inc., New York, and in Canada by Random House of Canada Limited, Toronto. Originally published in hardcover in the United States by Doubleday, a division of Random House, Inc., New York, in 2009.

Anchor Books and colophon are registered trademarks of Random House, Inc.

The Library of Congress has catalogued the Doubleday edition as follows:
Mezrich, Ben, 1969–
The accidental billionaires : the founding of Facebook, a tale of sex, money, genius and betrayal / Ben Mezrich.—1st ed.
p.cm.
Includes bibliographical references (pp. 259–60).
1. Zuckerberg, Mark, 1984– 2. Saverin, Eduardo. 3. Facebook (Firm).
4. Facebook (Electronic resource). 5. Webmasters—United States—Biography.
6. College students—United States—Biography.
7. Businesspeople—United States—Biography.
8. Online social networks. I. Title
HM742.M49 2009
006.7'54—dc22
2009417514

Anchor ISBN: 978-0-7679-3155-7

www.anchorbooks.com

Printed in the United States of America
10 9 8 7 6 5 4 3

TO TONYA,

THIS GEEK'S DREAM GIRL . . .

CONTENTS

THE
accidental billionaires

AUTHOR'S NOTE

The Accidental Billionaires is a dramatic, narrative account based on dozens of interviews, hundreds of sources, and thousands of pages of documents, including records from several court proceedings.

There are a number of different—and often contentious—opinions about some of the events that took place. Trying to paint a scene from the memories of dozens of sources—some direct witnesses, some indirect—can often lead to discrepancies. I re-created the scenes in the book based on the information I uncovered from documents and interviews, and my best judgment as to what version most closely fits the documentary record. Other scenes are written in a way that describes individual perceptions without endorsing them.

I have tried to keep the chronology as close to exact as possible. In some instances, details of settings and descriptions have been changed or imagined, and identifying details of certain people altered to protect their privacy. Other than the handful of public figures who populate this story, names and personal descriptions have been altered.

I do employ the technique of re-created dialogue. I have based this dialogue on the recollections of participants of the substance of conversations. Some of the conversations recounted in this book took place over long periods of time, in multiple locations, and thus some conversations and scenes were re-created and compressed. Rather than spread these conversations out, I sometimes set these scenes in likely settings.

I address sources more fully in the acknowledgments, but it is appropriate here to thank in particular Will McMullen for introducing me to Eduardo Saverin, without whom this story could not have been written. Mark Zuckerberg, as is his perfect right, declined to speak with me for this book despite numerous requests.

CHAPTER 1 | OCTOBER 2003

It was probably the third cocktail that did the trick. It was hard for Eduardo to tell for sure, because the three drinks had come in such rapid succession—the empty plastic cups were now stacked accordion style on the windowsill behind him—that he hadn't been able to gauge for certain when the change had occurred. But there was no denying it now, the evidence was all over him. The pleasantly warm flush to his normally sallow cheeks; the relaxed, almost rubbery way he leaned against the window—a stark contrast to his usual calcified, if slightly hunched posture; and most important of all, the easy smile on his face, something he'd practiced unsuccessfully in the mirror for two hours before he'd left his dorm room that evening. No doubt at all, the alcohol had taken effect, and Eduardo wasn't scared anymore. At the very least, he was no longer overwhelmed with the intense urge *to get the fuck out of there.*

To be sure, the room in front of him was intimidating: the immense crystal chandelier hanging from the arched, cathedral ceiling; the thick red velvet carpeting that seemed to bleed right out of the regal mahogany walls; the meandering, bifurcated staircase that snaked up toward the storied, ultrasecret, catacombed upper floors. Even the windowpanes behind Eduardo's head seemed treacherous, lit from behind by the flickering anger of a bonfire consuming most of the narrow courtyard outside, twists of flame licking at the ancient, pockmarked glass.

This was a terrifying place, especially for a kid like Eduardo. He hadn't grown up poor—he'd spent most of his childhood being shuttled between upper-middle-class communities in Brazil and Miami before matriculating at Harvard—but he was a complete stranger to the sort of old-world opulence this room represented. Even through the booze, Eduardo could feel the insecurities rumbling deep down in the pit of his stomach. He felt like a freshman all over again, stepping into Harvard Yard for the first time, wondering what the hell he was doing there, wondering how he could possibly belong in a place like that. *How he could possibly belong in a place like this.*

He shifted against the sill, scanning the crowd of young men that filled most of the cavernous room. A mob, really, bunched together around the pair of makeshift bars that had been set up specifically for the event. The bars themselves were fairly shoddy—wooden tables that were little more than slabs, starkly out of character in such an austere setting—but nobody noticed, because the bars were staffed by the only girls in the room; matching, bust-heavy blondes in low-cut black tops, brought in from one of the local all-female colleges to cater to the mob of young men.

The mob, in many ways, was even more frightening than the building itself. Eduardo couldn't tell for sure, but he guessed there had to be about two hundred of them—all male, all dressed in similar dark blazers and equally dark slacks. Sophomores, mostly; a mix of races, but there was something very similar about all the faces—

the smiles that seemed so much easier than Eduardo's, the confidence in those two hundred pairs of eyes—these kids weren't used to having to prove themselves. *They belonged.* For most of them, this party—this place—was just a formality.

Eduardo took a deep breath, wincing slightly at the bitter tinge to the air. The ash from the bonfire outside was making its way through the windowpanes, but he didn't move away from his perch against the sill, not yet. He wasn't ready yet.

Instead, he let his attention settle on the group of blazers closest to him—four kids of medium build. He didn't recognize any of them from his classes; two of the kids were blond and preppy-looking, like they'd just stepped off a train from Connecticut. The third was Asian, and seemed a little older, but it was hard to tell for sure. The fourth, however—African American and very polished-looking, from his grin to his perfectly coiffed hair—was definitely a senior.

Eduardo felt his back stiffen, and he glanced toward the black kid's tie. The color of the material was all the verification Eduardo needed. The kid was a senior, and it was time for Eduardo to make his move.

Eduardo straightened his shoulders and pushed off of the sill. He nodded at the two Connecticut kids and the Asian, but his attention remained focused on the older kid—and his solid black, uniquely decorated tie.

"Eduardo Saverin." Eduardo introduced himself, vigorously shaking the kid's hand. "Great to meet you."

The kid responded with his own name, Darron something, which Eduardo filed away in the back of his memory. The kid's name didn't really matter; the tie alone told him everything he needed to know. The purpose of this entire evening lay in the little white birds that speckled the solid black material. The tie designated him as a member of the Phoenix-S K; he was one of twenty or so hosts of the evening's affair, who were scattered among the two hundred sophomore men.

"Saverin. You're the one with the hedge fund, right?"

Eduardo blushed, but inside he was thrilled that the Phoenix member recognized his name. It was a bit of an exaggeration—he didn't have a hedge fund, he'd simply made some money investing with his brother during his sophomore summer—but he wasn't going to correct the mistake. If the Phoenix members were talking about him, if somehow they were impressed by what they'd heard—well, maybe he had a chance.

It was a heady thought, and Eduardo's heart started to beat a little harder as he tried to spread just the right amount of bullshit to keep the senior interested. More than any test he'd taken freshman or sophomore year, this moment was going to define his future. Eduardo knew what it would mean to gain entrance to the Phoenix—for his social status during his last two years of college, and for his future, whatever future he chose to chase.

Like the secret societies at Yale that had gotten so much press over the years, the Final Clubs were the barely kept secret soul of campus life at Harvard; housed in centuries-old mansions spread out across Cambridge, the eight all-male clubs had nurtured generations of world leaders, financial giants, and power brokers. Almost as important, membership in one of the eight clubs granted an instant social identity; each of the clubs had a different personality, from the ultra-exclusive Porcellian, the oldest club on campus, whose members had names like Roosevelt and Rockefeller, to the prepped-out Fly Club, which had spawned two presidents and a handful of billionaires, each of the clubs had its own distinct, and instantly defining, power. The Phoenix, for its part, wasn't the most prestigious of the clubs, but in many ways it was the social king of the hill; the austere building at 323 Mt. Auburn Street was the destination of choice on Friday and Saturday nights, and if you were a member of the Phoenix, not only were you a part of a century-old network, you also got to spend your weekends at the best parties on campus, surrounded by the hottest girls culled from schools all over the 02138 zip code.

"The hedge fund is a hobby, really," Eduardo humbly confided as the small group of blazers hung on his words. "We focus mostly on oil futures. See, I've always been obsessed with the weather, and I made a few good hurricane predictions that the rest of the market hadn't quite picked up on."

Eduardo knew he was walking a fine line, trying to minimize the geekiness of how he'd actually outguessed the oil market; he knew the Phoenix member wanted to hear about the three hundred thousand dollars Eduardo had made trading oil, not the nerdish obsession with meteorology that had made the trades possible. But Eduardo also wanted to show off a little; Darron's mention of his "hedge fund" only confirmed what Eduardo had already suspected, that the only reason he was in that room in the first place was his reputation as a budding businessman.

Hell, he knew he didn't have much else going for him. He wasn't an athlete, didn't come from a long line of legacies, and certainly wasn't burning up the social scene. He was gawky, his arms were a little too long for his body, and he only really relaxed when he drank. But still, he was there, in that room. A year late—most people were "punched" during the fall of their sophomore year, not as juniors like Eduardo—but he was there just the same.

The whole punch process had taken him by surprise. Just two nights before, Eduardo had been sitting at his desk in his dorm room, working on a twenty-page paper about some bizarre tribe that lived in the Amazonian rain forest, when an invitation had suddenly appeared under his door. It wasn't anything like a fairy-tale golden ticket—of the two hundred mostly sophomores who were invited to the first punch party, only twenty or so would emerge as new members of the Phoenix—but the moment was as thrilling to Eduardo as when he had opened his Harvard acceptance letter. He'd been hoping for a shot at one of the clubs since he'd gotten to Harvard, and now, finally, he'd gotten that shot.

Now it was just up to him—and, of course, the kids wearing the black, bird-covered ties. Each of the four punch events—like tonight's meet-and-greet cocktail party—was a sort of mass interview. After Eduardo and the rest of the invitees were sent home to their various dorms spread across the campus, the Phoenix members would convene in one of the secret rooms upstairs to deliberate their fates. After each event, a smaller and smaller percentage of the punched would get the next invitation—and slowly, the two hundred would be weeded down to twenty.

If Eduardo made the cut, his life would change. And if it took some creative "elaboration" of a summer spent analyzing barometric changes and predicting how those changes would affect oil distribution patterns—well, Eduardo wasn't above a little applied creativity.

"The real trick is figuring out how to turn three hundred thousand into three million." Eduardo grinned. "But that's the fun of hedge funds. You get to be real inventive."

He delved into the bullshit with full enthusiasm, carrying the whole group of blazers with him. He'd honed his bullshit skills over numerous prepunch parties as a freshman and sophomore; the trick was to forget that this was no longer a dry run—that this was the real thing. In his head, he tried to pretend he was back at one of those less important mixers, when he wasn't yet being judged, when he wasn't trying to end up on some all-important list. He could remember one, in particular, that had gone incredibly well; a Caribbean-themed party, with faux palm trees and sand on the floor. He tried to put himself back there—remembering the less imposing details of the decor, remembering how simple and easy the conversation had come. Within moments, he felt himself relaxing even more, allowing himself to become enrapt in his own story, the sound of his own voice.

He was back at that Caribbean party, down to the last detail. He remembered the reggae music bouncing off the walls, the sound of steel drums biting at his ears. He remembered the rum-based punch, the girls in flowered bikinis.

He even remembered the kid with the mop of curly hair who had been standing in a corner of the room, barely ten feet away from where he was now, watching his progress, trying to get up the nerve to follow his lead and approach one of the older Phoenix kids before it was too late. But the kid had never moved from the corner; in fact, his self-defeating awkwardness had been so palpable, it had acted like a force field, carving out an area of the room around him, a sort of reverse magnetism, pushing anyone nearby away.

Eduardo had felt a tinge of sympathy at the time—because he had recognized that kid with the curly hair—and because there was no way in hell a kid like that was ever going to get into the Phoenix. A kid like that had no business punching any of the Final Clubs—God only knew what he had been doing there at the prepunch party in the first place. Harvard had plenty of little niches for kids like that; computer labs, chess guilds, dozens of underground organizations and hobbies catering to every imaginable twist of social impairment. One look at the kid, and it had been obvious to Eduardo that he didn't know the first thing about the sort of social networking one had to master to get into a club like the Phoenix.

But then, as now, Eduardo had been too busy chasing his dream to spend much time thinking about the awkward kid in the corner.

Certainly, he had no way of knowing, then or now, that the kid with the curly hair was one day going to take the entire concept of a social network and turn it on its head. That one day, the kid with the curly hair struggling through that prepunch party was going to change Eduardo's life more than any Final Club ever could.

CHAPTER 2 | HARVARD YARD

Ten minutes past one in the morning, and something had gone terribly wrong with the decorations. It wasn't just that the ribbons of white- and blue-colored crepe paper attached to the walls had started to droop—one of them bowing so low that its taffeta-like curls threatened to overwhelm the oversize punch bowl perched below—but now the brightly designed decorative posters that covered much of the bare space between the crepe paper had also begun to unhinge and drop to the floor at an alarming rate. In some areas, the beige carpet had almost vanished beneath piles of glossy computer-printed pages.

On closer inspection, the catastrophe of the decorations made more sense; the peeling strips of packing tape that held the colored posters and crepe-paper ribbons in place were clearly visible, and what's more, a sheen of condensation was slowly working the strips of tape free as the heat from the overworked radiators that lined the walls played havoc with the hastily constructed ambience.

The heat was necessary, of course, because it was New

England in October. The banner hanging from the ceiling above the dying posters was all warmth—ALPHA EPSILON PI, MEET AND GREET, 2003—but there was no way a banner could compete with the ice that had begun to form on the oversize windows lining the back wall of the cavernous lecture room. All in all, the decorating committee had done what they could with the room—normally home to numerous philosophy and history classes, lodged as it was deep into the fifth floor of an aging building in Harvard Yard. They'd carted away the row upon row of scuffed wooden chairs and dilapidated desks, tried to cover up the bland, chipped walls with posters and crepe, and put up the banner, concealing most of the ugly, oversize fluorescent ceiling lights. Topping it all off, there was the coup de grâce; an iPod player attached to two enormous and expensive-looking speakers set on the little stage at the head of the room, where the professor's lectern usually stood.

Ten minutes to one in the morning, and the iPod was churning away, filling the air with a mixture of pop and anachronistic folk rock—either the result of a schizophrenic's playlist or some bickering committee members' poorly thought-out compromise. Even so, the music wasn't half bad—and the speakers were a minor coup brought about by whoever was in charge of the entertainment. A previous year's shindig had featured a color television in the back corner of the classroom, hooked up to a borrowed DVD of Niagara Falls playing on an infinite loop. No matter that Niagara Falls had nothing even remotely to do with Alpha Epsilon Pi or Harvard; the sound of running water had somehow seemed party appropriate, and it hadn't cost the committee a dime.

The speaker system was an upgrade—as were the peeling posters. The party, on the other hand, was par for the course.

Eduardo stood beneath the banner, thin slacks hanging down

over his storkish legs, an oxford shirt buttoned all the way up to his throat. Surrounding him were four similarly attired guys, mostly juniors and sophomores. Together, the small group made up a full third of the party. Somewhere, on the other side of the room, there were two or three girls in the mix. One of them had even dared to wear a skirt to the event—although she'd chosen to wear it over thick gray leggings, out of respect for the weather.

It wasn't exactly a scene from *Animal House,* but then, underground fraternity life at Harvard was a far cry from the Greek bacchanalia one might find at other colleges. And Epsilon Pi wasn't exactly the pearl of the undergrounds; as the reigning Jewish frat on campus, its membership was more notorious for its combined grade-point average than its party proclivities. This reputation had nothing to do with its nominal religious leanings; the truly pious Jews, the ones who kept kosher and dated only within the tribe, joined Hillel House, which had its own building on campus and sported a true endowment, not to mention both male and female members. Epsilon Pi was for the secular kids, the ones whose last names were their most recognizably Jewish feature. To the Epsilon Pi kids, a Jewish girlfriend might be nice because it would make Mom and Dad happy. But, in reality, an Asian girlfriend was much more likely.

Which was exactly what Eduardo was explaining to the frat brothers surrounding him—a topic of conversation they'd revisited fairly frequently, because it hinged on a philosophy they could all get behind.

"It's not that guys like me are generally attracted to Asian girls," Eduardo commented, between sips of punch. "It's that Asian girls are generally attracted to guys like me. And if I'm trying to optimize my chances of scoring with the hottest girl possible, I've got to stock my pond with the type of girls who are the most likely to be interested."

The other kids nodded, appreciating his logic. In the past, they'd taken this simple equation and elaborated it into a much more complex algorithm to try to explain the connection between Jewish guys and Asian girls, but tonight they just let it remain simple, perhaps out of respect for the music, which was now reverberating so loudly through the expensive speakers that it was hard to engage in any complex thought.

"Although at the moment"—Eduardo grimaced as he glanced toward the girl in the skirt-leggings combination—"this pond's running a little dry."

Again, agreement all around, but it wasn't like any of his four frat mates were going to do anything about the situation. The kid to Eduardo's right was five foot six and pudgy; he was also on the Harvard chess team and spoke six languages fluently, but none of that seemed to help when it came to communicating with girls. The kid next to him drew a cartoon strip for the *Crimson*—and spent most of his free time playing RPG video games in the student lounge above the Leverett House dining hall. The cartoonist's roommate, standing next to him, was well over six feet tall; but instead of basketball, he'd chosen fencing as a high school student at a mostly Jewish prep school; he was good with an épée, which was about as useful when it came to picking up girls as it was in any other aspect of modern life. If eighteenth-century pirates ever attacked a hot girl's dorm room, he was ready, but otherwise he was pretty much useless.

The fourth kid, standing directly across from Eduardo, had also fenced—at Exeter—but he wasn't built anything like the tall kid to his left. He was a bit on the gawky side, like Eduardo, though his legs and arms were more proportionate to his slim, not entirely unathletic frame. He was wearing cargo shorts instead of slacks, sandals with no socks. He had a prominent nose, a mop of curly blondish brown hair, and light blue eyes. There was something playful about those eyes—but that was where any sense of natural emotion or readability ended. His narrow face was otherwise devoid of any expression at all. And his posture, his general aura—the way he seemed closed in on himself, even while engaged in a group dynamic, even here, in the safety of his own fraternity—was almost painfully awkward.

His name was Mark Zuckerberg, he was a sophomore, and although Eduardo had spent a fair amount of time at various Epsilon Pi events with him, along with at least one prepunch Phoenix event that Eduardo could remember, he still barely knew the kid. Mark's reputation, however, definitely preceded him: a computer science major who lived in Eliot House, Mark had grown up in the upper-middle class town of Dobbs Ferry, New York, the son of a dentist and a psychiatrist. In high school, he'd supposedly been some sort of master hacker—so good at breaking into computer systems that he'd ended up on some random FBI list somewhere, or so the story went. Whether or not that was true, Mark was certainly a computer genius. He had also made a name for himself at Exeter when, after he had honed his coding skills creating a computerized version of the game Risk, he and a buddy had created a software program called Synapse, a plug-in for MP3 players that allowed the players to "learn" a user's preferences and create tailored playlists based on that information. Mark had posted Synapse as a free download on the Web—and almost immediately, major companies came calling, trying to buy Mark's creation. Rumor was, Microsoft had offered Mark between one and two million dollars to go work for them—and amazingly, Mark had turned them down.

Eduardo wasn't an expert on computers, and he knew very little about hacking, but business ran in his family, and the idea that someone would turn down a million dollars was fascinating—and a little bit appalling—to him. Which made Mark more of an enigma than even his awkwardness implied. An enigma—and obviously a genius. He'd followed Synapse up with a program he'd written at Harvard, something called Course Match that allowed Harvard kids to see what classes other kids had signed up for; Eduardo had checked it out himself once or twice, trying to track down random hot girls he'd met in the dining hall, to little avail. But the program was good enough to get a pretty big following; most of the campus appreciated Course Match—if not the kid who'd created it.

As the three other frat brothers moved off toward the punch bowl for a refill, Eduardo took the opportunity to study the moppet-haired sophomore a little closer. Eduardo had always prided himself on his ability to get to the core of other people's personalities—it was something his father had taught him, a way of getting a step ahead in the world of business. For his father, business was everything; the son of wealthy immigrants who had barely escaped the Holocaust to Brazil during World War II, his father had raised Eduardo in the sometimes harsh light of survivors; he came from a long line of businessmen who knew how important it was to succeed, whatever one's circumstances. And Brazil was only the beginning; the Saverin family had almost just as hastily been forced to relocate to Miami when Eduardo was thirteen—when it was discovered that Eduardo's name had ended up on a kidnap list because of his father's financial success.

By junior high, Eduardo had found himself adrift in a strange new world, struggling to learn a new language—English—and a new culture—Miami—at the same time. So he didn't know computers, but he understood, completely, what it was like being the awkward outsider; being different, whatever the reason.

Mark Zuckerberg, from the looks of him, was obviously different. Maybe it was just that he was so damn smart, he didn't fit in, even here, among his peers. Among his own kind: not Jewish, per se, but kids like him. Geeky kids who made algorithms out of fetishes, who had nothing better to do on a Friday night than hang out in a classroom filled with crepe paper and colored posters, talking about girls they weren't actually getting.

"This is fun," Mark finally said, breaking the silence. There was almost zero inflection in his voice, and it was impossible for Eduardo to guess what emotion—if any—he was trying to convey.

"Yeah," Eduardo responded. "At least the punch has rum in it this year. Last time, I think it was Capri Sun. They went all out this time around."

Mark coughed, then reached out toward one of the crepe-paper ribbons, touching the closest twist of material. The packing tape unhinged, and the ribbon drifted toward the floor, landing on his Adidas sandal. He looked at Eduardo.

"Welcome to the jungle."

Eduardo grinned, despite the fact that he still couldn't be sure from Mark's monotone delivery if the kid was joking or not. But he was getting the sense that there was something really anarchistic going on behind the kid's blue eyes. He seemed to be taking everything in around him, even here, in a place with so little stimulus to grasp onto. Maybe he really was the genius everyone thought he was. Eduardo had the abrupt feeling that this was someone he wanted to befriend, to get to know better. Anyone who'd turned down a million dollars at seventeen was probably heading somewhere.

"I have a feeling this is gonna break up in a few minutes," Eduardo said. "I'm heading back to the river—Eliot House. What house are you in again?"

"Kirkland," Mark responded. He jerked his head toward the exit, on the other side of the stage. Eduardo glanced at their other friends, still at the punch bowl; they were all quad kids, so they'd be going in a different direction when the party ended. It was as good an opportunity as any to get to know the awkward computer genius. Eduardo nodded, then followed Mark through the sparse crowd.

"If you want," Eduardo offered as they wound their way around the stage, "there's a party on my floor we could check out. It's gonna suck, but certainly no worse than this."

Mark shrugged. They'd both been at Harvard long enough to know what to expect from a dorm party; fifty dudes and about three girls jammed into a small, coffinlike box of a room, while someone tried to figure out how to tap an illicit keg of really cheap beer.

"Why not," Mark responded, over his shoulder. "I've got a problem set due tomorrow, but I'm better at logarithms drunk than sober."

A few minutes later, they had pushed their way out of the lecture room and into the cement stairwell that descended to the ground floor. They took the steps in silence, bursting out through a pair of double doors into the tree-lined quiet of Harvard Yard. A stiff, crisp breeze whipped through the thin material of Eduardo's shirt. He jammed his hands into the deep pockets of his slacks and started forward down the paved path that led through the center of the Yard. It was a good ten-minute walk to the houses on the river, where both he and Mark lived.

"Shit, it's fucking ten degrees out here."

"More like forty," Mark replied.

"I'm from Miami. It's ten degrees to me."

"Then maybe we should run."

Mark took off at a slow jog. Eduardo followed suit, breathing hard as he caught up to his new friend. They were side by side as they moved past the impressive stone steps that led up to the pillared entrance to Widener Library. Eduardo had spent many evenings lost in the stacks of Widener—poring through the works of economic theorists such as Adam Smith, John Mills, even Galbraith. Even after one in the morning, the library was still open; warm orange light from inside the marbled lobby splashed out through the glass doors, casting long shadows down the magnificent steps.

"Senior year," Eduardo huffed as they skirted the bottom stone step on their way to the iron gate that led out of the Yard and off into Cambridge, "I'm going to have sex in those stacks. I swear, it's gonna happen."

It was an old Harvard tradition—something you were supposed to do before you graduated. The truth was, only a handful of kids had ever actually achieved the mission. Though the automated stacks—vast bookshelves on automatic, wheeled tracks—were labyrinthine and descended many floors below the massive building, there were always students and staff lurking through those narrow passageways; finding a spot isolated enough to do the deed would be quite a feat. And finding a girl who was willing to attempt to continue the tradition was even more unlikely.

"Baby steps," Mark responded. "Maybe you should try getting a girl back to your dorm, first."

Eduardo winced, then grinned again. He was starting to like this kid's caustic sense of humor.

"Things aren't that bad. I'm punching the Phoenix."

Mark glanced at him as they turned the corner and headed along the side of the great library.

"Congratulations."

There it was again, zero inflection. But Eduardo could tell from the little flash in Mark's eyes that he was impressed, and more than a little envious. That was the reaction Eduardo had learned to expect when he mentioned the punch process he was going through. The truth was, he'd been letting it slip to everyone he knew that he was getting closer and closer to becoming a member of the Phoenix. He'd been through three punch events already; there was a very good chance, now, that he'd go the distance. And maybe, just maybe, events like the Alpha Epsilon Pi party they'd just survived would be a thing of the past.

"Well, if I get in, maybe I can put your name on the list. For next year. You could punch as a junior."

Mark paused again. Maybe he was catching his breath. More likely, he was digesting the information. There was something very computer-like about the way he spoke; input in, then input out.

"That would be—interesting."

"If you get to know some of the other members, you'll have a good shot. I'm sure a lot of them used your Course Match program."

Eduardo knew, as he said it, how foolish the idea sounded. Phoenix members weren't going to get excited about this awkward kid because of some computer program. You didn't get popular by writing computer code. A computer program couldn't get you laid. You got popular—and sometimes laid—by going to parties, hanging out with pretty girls.

Eduardo hadn't gotten that far yet, but last night he had received that all-important fourth punch invitation. In one week, next Friday night, there was a banquet at the nearby Hyatt hotel, then an after-party at the Phoenix. It was a big night, perhaps the final big punch event before new members were initiated. The invitation had "suggested" that Eduardo bring a date to the dinner; he'd heard from classmates that in fact the members would be judging the punches on the quality of the women they brought with them. The better-looking their dates, the more likely it was that they'd get through to the final punch round.

After receiving the letter, Eduardo had wondered how the hell he was going to get a date—an impressive one, at that—on such short notice. It wasn't like the girls were breaking down his dorm-room door.

So Eduardo had been forced to take matters into his own hands. At nine A.M. that morning, in the Eliot dining hall, he had walked right up to the hottest girl he knew—Marsha, blond, buxom, in reality an econ major but she looked like a psychology major. She was a good two inches taller than Eduardo, and had a strange predisposition toward eighties-style hair scrunchies, but she was beautiful, in a Northeast prep-school sort of way. In short, she was perfect for the punch event.

To Eduardo's surprise, she'd said yes. Eduardo had immediately realized—it was the Phoenix, it wasn't about Eduardo—it was about going to a Final Club dinner. Which bolstered everything Eduardo already believed about the Final Clubs. Not only were they a powerful social network, but their exclusive nature gave their members instant status—the ability to attract the coolest, hottest, best. He had no illusions that Marsha was going to join him in the Widener stacks after the event—but at the very least, if enough alcohol was involved, she might let him walk her home. Even if she brushed him off at the door to her room with a little kiss, that would be further than he'd gotten in four months.

As they reached the back corner of the library and jogged out from under the long shadow of the building's archaic, stone pillars, Mark shot him another unreadable glance.

"Was it everything you hoped it would be?"

Was he talking about the library? The party they had just left? The Jewish fraternity? The Phoenix? Two geeky kids running across

Harvard Yard, one in a buttoned-up oxford shirt, the other in cargo shorts, freezing to death while they tried to get to some lousy dorm party?

For guys like Eduardo and Mark, was college life supposed to get any better than this?

CHAPTER 3 | ON THE CHARLES

Five A.M.

A desolate stretch of the Charles River, a quarter-mile serpent's twist of glassy greenish blue, braced by the arched stone Weeks Footbridge on one side and the concrete, multi-lane Mass Ave. Bridge on the other. A frigid glade of water winding beneath a gray-on-gray canopy of fog, hanging low and heavy, air so thick with moisture it was hard to tell where the river ended and the sky began.

Dead silence, a moment frozen in time, a single paragraph on a single page in a book that spanned three centuries of pregnant, frozen moments like this. Dead silence—and then, the slightest of noises: the sound of two knifelike oars dipped expertly into that frigid glade, pivoting beneath the swirl of greenish blue, levering backward in a perfect and complex marriage of mechanics and art.

A second later, a two-man skiff slid out from under the shadow of Weeks Bridge, its phallic, fiberglass body slicing down the center of the curving river like a diamond-edged blade carving its way across a windowpane. The motion of the craft was so smooth, the boat almost seemed a part of the water itself; the curved, fiberglass hull of the skiff seemed to bleed out of the green-blue water, its forward motion so pure it produced almost no wake.

One look at the skiff, the way the oars pierced the surface of the Charles in perfect rhythm, the way the boat glided across the water—and it was obvious that the two young men guiding the elegant device had spent years perfecting their art. One look at the two young men themselves, and it was equally obvious that it was more than just practice that had brought them to such a level of perfection.

From the shore, the two rowers looked like robots: exact replicas of each other, from their sandy-colored, full heads of hair to their chiseled all-American facial features. Like the progress of their craft, physically, they were near perfect. Muscles rippling beneath gray Harvard Crew sweatshirts, bodies built long and lithe, the two young men were easily six foot five inches tall apiece; impressive presences made more so by the fact that they were truly identical, from the piercing blue color of their eyes to the fiercely determined expressions on their matinee idol faces.

Technically, the Winklevoss brothers were "mirror" identical twins—the result of a single ovarian egg that had flipped open like two pages of a magazine. Tyler Winklevoss, at the front of the two-man skiff, was right-handed—and the more logical, serious-minded of the brothers. Cameron Winklevoss, at the rear of the boat, was left-handed; he was the more creative and artistic of the two.

At the moment, however, their personalities had merged; they didn't speak as they worked the oars—they didn't communicate at all, verbally or otherwise, as they effortlessly pushed the boat forward down the Charles. Their concentration was almost inhuman, the result of years spent honing their innate abilities under various coaches at

Harvard, and before that, in Greenwich, Connecticut, where the twins had grown up. In many ways, their hard work had already paid off; as college seniors, they were on track to make the Olympic rowing team. At Harvard, they were among the best of the best; crowned junior national champions the year before, they had led the Crimson to numerous crew-team victories, and they currently sat atop the Ivy League standings in any number of rowing categories.

But none of that mattered to the Winklevoss twins as they powered their boat across the frigid water. They had been out on the Charles since four, piloting back and forth between the two bridges—and they would continue their silent vigil for at least the next two hours. They would pull those oars until they were both near exhaustion, until the rest of the campus finally came alive—until bright yellow ribbons of sunlight finally began to break through that gray-on-gray fog.

▶ ▶ ▶

Three hours later, Tyler could still feel the river resonating beneath him as he dropped into a chair next to Cameron at the head of a long, scuffed wooden table in a back corner of the dining hall at Pforzheimer House. The hall was fairly modern and vast, a brightly lit, rectangular room with high ceilings, containing more than a dozen long tables; most of the tables were crowded with students, as it was already deep into the breakfast session.

Pforzheimer House was one of the newest of the Harvard undergraduate houses—"new" being a relative concept on a campus that was more than three hundred years old—and one of the biggest, home to about a hundred and fifty sophomores, juniors, and seniors. Freshmen lived in Harvard Yard; at the end of the freshman year, students entered a lottery system to find out where they'd spend the rest of their Harvard career—and Pforzheimer wasn't exactly at the top of anyone's wish list, located as it was in the center of "the Quad," a pretty little quadrangle of buildings surrounding a wide expanse of rolling grass—located precisely in the middle of nowhere. The Quad had been part of the university's expansion deep into Cambridge; ostensibly, to deal with overcrowding, but more likely simply to make better use of the huge financial endowment the university had amassed.

The Quad wasn't exactly Siberia, but to the students who were "quaded" at the end of freshman year, it certainly felt like they were about to be sent to some sort of gulag. The Quad houses were a good twenty-minute walk from Harvard Yard, where most of the undergraduate classes took place. For Tyler and Cameron, ending up in the Quad had been an even more difficult sentence; after the hike to the Yard, it was another ten-minute slog over to the river, where the Harvard boathouse squatted alongside the better-known Harvard houses: Eliot, Kirkland, Leverett, Mather, Lowell, Adams, Dunster, and Quincy.

Over there, the houses were known by their names. Out here, it was just the Quad.

Tyler glanced over at Cameron, who was leaning over a red plastic tray overflowing with breakfast items. A mountain range of scrambled eggs towered over foothills of breakfast potatoes, buttered toast, and fresh fruit, enough carbs to power an SUV—or a six-foot, five-inch rowing star. Tyler watched Cameron attacking the eggs, and could tell that his brother was nearly as worn out as he was. They'd been going full steam for the past few weeks—and not just out on the river, but also in their classes—and it was all starting to take its toll. Getting up every morning at four, heading down to the river. Then classes, homework. Then back down to the river for more training, weights, running. The life of a college athlete was hard; there were some days when it seemed like all they did was row, eat, and sometimes sleep.

Tyler shifted his gaze from Cameron and the scrambled eggs to the kid sitting across from them at the table. Divya Narendra was mostly hidden behind a copy of the *Crimson,* the school newspaper, which he was holding open in front of him with both hands. There was an untouched bowl of oatmeal beneath the newspaper, and Tyler was pretty sure that if Divya didn't put down the paper soon enough, Cameron would probably get that as well. If Tyler hadn't already finished a tray almost twice as loaded as Cameron's before he'd joined them at the rear table, he'd have taken the oatmeal down himself.

Divya wasn't an athlete like they were, but he certainly understood their passion and work ethic; he was as sharp a kid as Tyler had ever met, and together the three of them had been working pretty intensely on a somewhat secret project for quite some time. A sort of side venture in their lives, one that had slowly begun to take on more import—ironically—the busier their lives became.

Tyler cleared his throat, then waited for Divya to put the paper down so they could get started. Divya held up a finger, asking for a minute; Tyler rolled his eyes, frustrated. As he did so, his attention drifted to the table directly behind Divya. A group of girls kept glancing back at him and Cameron. When he looked right at them, they quickly looked away.

It was something Tyler had grown pretty used to, because it happened all the time. Hell, he and Cameron *were* identical twins. He knew that was something unusual—there was a slight freak-show element to it. But here, at Harvard, it was even more than that. They were on track to becoming Olympic athletes—and still, that was only part of it as well. Tyler and Cameron had a certain status on campus, a status that started with them being premier athletes—but carried over into something else.

The turning point, of course, was easy for Tyler to pinpoint. He and his brother had become members of the Porcellian Club during their junior year. That they had been punched as juniors was pretty unusual; not only was the Porcellian the most prestigious, secretive, and oldest Final Club on campus, but it was also the smallest in terms of number of members and new punches—and it was especially rare for students to get punched for the Porc a year late.

Tyler was pretty sure the club had waited the extra year to bring them in because of their background. Most members of the Porc had names with hundred year histories at Harvard. Although Tyler and Cameron's father was immensely wealthy, he'd made his money himself, building a highly successful consulting company from the ground up. Tyler and Cameron weren't from old money—but they certainly were from money. At the Fly or the Phoenix, that would have been enough. At the Porc, there had to be more.

The Porc, after all, wasn't a social institution like the Phoenix. For one thing, women were not allowed inside the club. On a member's wedding day, he could bring his wife on a tour of the building; then, during his twenty-fifth reunion, he could bring her again. And that was it. Only the famed Bicycle Room—a preparty hot spot, adjacent to the club proper—was accessible to nonmembers and coeds.

The Porc wasn't about parties or about getting laid like the other clubs on campus. It was about the future. It was about status—the sort of status that got you stared at in the dining hall, in the classrooms, when you were walking across Harvard Yard. The Porc wasn't a social club—it was serious business.

Which was something Tyler could appreciate. Serious business—after all, that's why he and his brother were meeting with Divya that morning in the dining hall, an hour later than they usually ate breakfast. *Serious fucking business.*

Tyler turned his attention away from the blushing girls at the next table, then grabbed a half-eaten apple off of his brother's tray. Before his brother could protest, he tossed the apple in a high arc, landing it in the center of Divya's bowl of oatmeal. The oatmeal splashed upward, soaking the newspaper with globs of thick white goo.

Divya paused, then carefully folded the destroyed newspaper and placed it on the table next to the bowl.

"Why do you read that rag?" Tyler asked, grinning at his friend. "It's a complete waste of time."

"I like to know what my fellow students are up to," Divya responded. "I think it's important to keep a finger on the pulse of the student body. One day we're going to launch this freaking company, and then this 'rag' is going to be real important to us, don't you think?"

Tyler shrugged, but he knew Divya was right. Divya was usually right. Which was the main reason Tyler and Cameron had partnered with him in the first place. They had been meeting like this, once a week, sometimes more often, since December of 2002. *Almost two full years.*

"Well, we're not going to launch *anything* unless we find someone to replace Victor," Cameron interrupted, around a mouthful of eggs. "That's for damn sure."

"He's really out?" Tyler asked.

"Yep," Divya responded. "He says he's got too much on his plate, he can't put any more time into this. We need a new programmer. And it's gonna be hard to find someone as good as Victor."

Tyler sighed. Two full years—and it seemed like they were no closer to launching than they were when they had first started. Victor Gua had been a great asset—a computer whiz who'd understood what they were trying to build. But he'd been unable to finish the site, and now he was gone.

If only Tyler, Cameron, or Divya had the necessary computer background to get the thing off and running—Christ, Tyler knew in his soul that the company was going to be a huge success. It was such an amazing idea—something Divya had initially come up with, then he and Cameron had helped hone into what they all humbly considered pure genius.

The project was called the Harvard Connection, and it was a Web site that was going to change life on campus—if they could only get someone to write the computer code that would make it work. The central idea was simple: put Harvard's social life online, make a site where guys like Tyler and Cameron—who spent all their time rowing, eating, and sleeping—could meet up with girls—like the ones stealing glances at them from the next table over—without all the inefficient, time-wasting, wandering around campus that real life usually necessitated.

As members of the Harvard elite, Tyler and Cameron were in a unique position to see how flawed Harvard's social scene was. Eligible guys—like them—never had the opportunity to meet enough choice girls, because they were too busy doing the things that made them such hot properties on campus. A Web site geared toward socializing could fix that problem, could create a fluid environment where girls and guys could meet.

The Harvard Connection would fulfill a need in what was mostly a stagnant social scene. Right now, if you rowed crew or played baseball or football—that's all you ever did. The only girls you ever met were the ones who hung around the river, or the baseball field, or the gridiron. If you lived in the Quad, Quad girls were all you'd ever have access to. Sure, you could drop the "H-bomb" on anyone within range—meaning, you could use your Harvard Male status to bring down interested parties within your proximity—but a site like the Harvard Connection would vastly increase your range.

Simple, perfect, fulfilling a need. The site would have two sections, dating and connecting. And once it succeeded at Harvard, Tyler and Cameron foresaw the site moving to other colleges, maybe throughout the Ivies. Every school had its own version of the H-bomb, after all.

The only flaw in their business plan was, they didn't have any way in hell of making the site without the help of a real computer genius. Tyler and Cameron had taught themselves HTML while in high school—but they weren't good enough to build something like this. Truth was, they needed a real geek to make their social site work. Not just someone smart—someone who got what they were trying to do. The Harvard Connection was going to be something Harvard kids actually used, every weekend, an addition to their social routines. You'd shower, shave, make some calls, then check out the Connection and see who'd been checking *you* out.

"Victor says he can find us some names," Divya continued as he shook the newspaper over the oatmeal bowl, trying to dry it off. "Some kids from his computer science classes. We can start interviewing people, throw the word around that we're looking for someone."

"I can ask around the Porc," Cameron added. "I mean, nobody there's gonna know much about computers, but maybe someone's got a younger brother."

Great, Tyler thought, next they'd be posting a wanted add in the science center and hanging around the computer labs. He watched Divya working at the newspaper, and despite his frustration, he had to smile. Divya was such a polished guy, the son of two Indian doctors from Bayside, Queens, who'd followed his older brother into Harvard. He was always well dressed, well coiffed, well spoken. Nobody would have guessed that he was a genius on the electric guitar—specifically, a technical master of heavy-metal riffs. In public, he was so freaking dapper. He even liked to keep his newspaper clean.

While he watched Divya and the newspaper, Tyler's gaze inadvertently shifted back to the table of girls behind his friend. The tallest of the group—a brunette with striking brown eyes, wearing a low-cut tank top under a carefully torn Harvard Athletics sweatshirt—was now looking right at him, smiling over a purposefully revealed slice of tan shoulder. Tyler couldn't help but smile back at her.

Divya coughed, interrupting Tyler's thoughts.

"I highly doubt she's interested in HTML code."

"Couldn't hurt to ask," Tyler responded as he winked at the brunette. Then he rose from the table. Their meeting had been short—but until they found themselves a new Victor, there wasn't much else that they could do. He started toward the group of girls, then paused to grin at his Indian friend and his oatmeal-covered newspaper.

"One thing I know for sure—you're not going to find us a computer programmer in the fucking *Crimson*."

CHAPTER 4 | CANNIBALISTIC CHICKENS

Eduardo pushed open the huge double doors as quietly as he could and slid into the back of the enormous lecture hall. The lecture was already in full swing; down at the bottom of the movie-theater-style room, on a raised stage that was backlit by a handful of industrial-size spotlights, a rotund little man in a tweed sport coat bounced up and down behind a massive oak lectern. The man was fully energized, his round little cheeks bright red with passion. His spindly arms jerked up and down, and every few minutes he slapped them against the lectern, sending a gunshotlike *pop* through the speakers that hung from the hall's ridiculously high ceilings. Then he'd gesture over his shoulder, where behind him, spread across a ten-foot-high blackboard, hung a full colored map that looked like a cross between something from a Tolkien book and something that might have hung in FDR's war room.

Eduardo had no idea what class this was, or what this lecture was about. He didn't recognize the professor, but that wasn't unusual; there were so many professors, teaching fellows, and senior tutors at Harvard, one couldn't possibly be expected to keep them all straight. He could tell from the size of the room—and the fact that the three-hundred-seat theater was near full—that it was some sort of Core requirement. Because only Core classes were this big—as they were mandatory, what students like Eduardo and Mark considered necessary evils of Harvard life.

The Core at Harvard was more than a requirement—it was also what the school considered a philosophy. The idea was that every student had to devote at least a quarter of their class time to courses that were designed to create a "rounded" scholar. The Core categories were foreign cultures, historical study, literature, moral reasoning, quantitative reasoning, science, and social analysis. The idea seemed sound; but in practice, the Core didn't come close to living up to its lofty ideals. Because at their heart, the Core classes catered to the lowest common denominator, as nobody took a Core course because they were actually interested in the subject. So instead of deep, scholarly courses on history and the arts, you had classes such as Folklore and Mythology—or as it was affectionately known by the kids who slept through its vast lectures, "Greeks for Geeks"; a simple intro to physics—"Physics for Poets." And a half-dozen bizarre anthropology courses that had little or no relevance to the real world. Because of the Core, nearly every Harvard graduate had taken at least one course that dealt with the Yanomamö, the "fierce people" of the Amazonian rain forest, a bizarre little tribe that still lived like they were in the Stone Age. A Harvard grad didn't need to know much politics or math; but ask about the Yanomamö, and any grad could tell you that they were very fierce—that they often fought each other with big long sticks and engaged in strange piercing rituals that were even more disturbing than those engaged in by the kids who hung out in the skateboarding pit in the center of Harvard Square.

From the back of the vast room, Eduardo watched the professor hop about behind the lectern, catching an odd word or phrase from the echoing sound system up above. From what he could tell, this particular Core class had something to do with history or philosophy; on closer inspection, the map behind the prof looked to be Europe sometime in the past three hundred years—but that didn't really clear things up. Eduardo doubted the class had anything to do with the Yanomamö, but at Harvard, you couldn't be sure.

This particular morning, he wasn't there to get himself a little more "well rounded." He was on a mission of a very different nature.

He scanned the room, using one hand to shield his eyes from the immense spots on the stage, which seemed to be aimed in exactly the wrong direction for what they'd been designed to do. His other hand was occupied; cradled beneath his left arm was a bulky crate, covered by a large blue towel. The crate was heavy, and Eduardo was very careful not to jostle the damn thing as he searched the rows of students for his quarry.

It took him a few minutes to locate Mark, sitting by himself three rows from the very back of the room. Mark had his sandaled feet up on the seat in front of him, which was empty, and a notebook spread open on his lap. He didn't seem to be taking any notes. In fact, he didn't seem to be awake at all; his eyes were closed, his head mostly covered by the oversize hood of the fleece he almost always wore, and his hands were jammed deep into the pockets of his jeans.

Eduardo grinned to himself; in a matter of a few short weeks, he and Mark had become close friends. Even though they lived in different houses and had different majors, Eduardo felt that they had a similar spirit—and he'd begun to notice an almost strange feeling that they were supposed to be friends, even before they were. In that short time, he'd grown to really like Mark, had begun to think of him like a real brother, not just someone who shared a Jewish frat, and he was pretty sure Mark felt the same way about him.

Still grinning, Eduardo quietly worked his way down the aisle to Mark's row. He stepped over the extended legs of a sleeping junior whom he barely recognized from one of his economics seminars, then pushed past a pair of sophomore girls who were both busy listening to an MP3 player stashed in the bag between them. Then he plopped down in the empty seat next to Mark, carefully placing the covered crate on the floor in front of his knees.

Mark opened his eyes, saw Eduardo sitting next to him—and then slowly turned his attention to the crate on the floor.

"Oh, shit."

"Yeah," Eduardo replied.

"That's not—"

"Yes, it is."

Mark whistled low, then leaned forward and lifted a corner of the blanket.

Instantaneously, the live chicken inside the corrugated milk crate started squawking at full volume. Feathers flew out of the crate, pluming upward, then raining down around Eduardo and Mark and anyone else within a five-yard radius. Kids in the rows in front of and behind where they were sitting gaped at them. Within a second, everyone in their part of the lecture hall was staring at them, a mixture of shock and amusement on their faces.

Eduardo's cheeks turned bright red and he quickly grabbed the towel and yanked it closed over the crate. Slowly, the bird quieted down. Eduardo glanced down to the stage—but the professor was still rambling on about Britons and Vikings and whoever the hell else ran around in that time period. Because of the overwhelming sound system, he hadn't noticed the commotion—thank God.

"That's great," Mark commented, grinning at the crate. "I really like your new friend. He's a much better conversationalist than you are."

"It's not great!" Eduardo hissed, ignoring Mark's jab. "This chicken is a pain in the ass. And it's caused me a whole shitload of trouble."

Mark just kept on grinning. To be fair, the situation was actually quite comical, when you looked at it from the outside. The chicken was part of Eduardo's Phoenix initiation; he had been instructed to keep it with him at all times, to carry it with him everywhere, day and night, to every class, dining hall, and dorm room he visited. Hell, he had to sleep with the damn thing. For five whole days, his only job had been to keep that chicken alive.

And for the first few days, everything had gone swimmingly. The chicken had seemed happy, and none of his teachers had been the wiser. He'd avoided most of his smaller seminars, feigning the flu. The dining halls and the dorm rooms had been easy; most of the other students on campus knew about the Final Club initiations, so nobody gave him much of a hard time. And what few authority figures he ran across in his daily routine were willing to turn a blind eye. Getting into a Final Club was a big deal, and everybody knew it.

But in the last two days of his initiation, things had gotten more complicated.

It had all gone downhill forty-eight hours earlier, when Eduardo had brought the chicken back to his dorm room in Eliot House after a long day of dodging classes. It had turned out that down the hall from Eduardo's room lived two kids who were members of the Porcellian Club; Eduardo had met them a few times, but since they traveled in such different circles, they'd never really gotten to know one another. Eduardo hadn't thought anything of it when the two kids saw him with the chicken. Nor did he bother hiding the fact that for dinner, he'd decided to feed the chicken some fried chicken he'd smuggled home from the dining hall.

It wasn't until twenty-four hours later, when the *Harvard Crimson* published an explosive exposé, that Eduardo had realized what had happened. That evening, after witnessing Eduardo feeding chicken to the chicken, the Porc kids had written an anonymous e-mail to an animal rights group called the United Poultry Concern. The e-mail, signed by someone calling herself "Jennifer"—the e-mail address read friendofthePorc@hotmail.com—accused the Phoenix of ordering its new members to torture and kill live chickens as part of its initiation. The United Poultry Concern had immediately contacted the Harvard administration, reaching as high up as President Larry Summers himself. An ad-board investigation was already under way—and the Phoenix was going to have to defend itself against accusations of animal cruelty—including forcing cannibalism on defenseless poultry.

All in all, Eduardo had to admit that it was a pretty good prank by the Porc kids—but it was a huge headache for the Phoenix. Thankfully, the Phoenix leadership hadn't traced the fiasco back to Eduardo yet—though even if they did, they'd hopefully see the humor in the situation.

Of course, Eduardo hadn't been ordered to torture and kill his chicken. Exactly the opposite, he'd been ordered to keep his chicken healthy and alive. Maybe feeding the chicken chicken was a mistake; how was he supposed to know what chickens ate? The thing hadn't come with a manual. Eduardo had gone to a Jewish prep school in Miami. What the hell did Jews know about chickens, other than the fact that they made pretty good soup?

The entire debacle had almost overshadowed the fact that Eduardo was nearly finished with his initiation period. In a few more days, he was going to be a full-fledged member of the Phoenix. If the chicken fiasco didn't end up getting him kicked out, pretty soon he'd be hanging out in the club every weekend, and his social life was going to change dramatically. Already, those changes had begun to take effect.

He leaned toward Mark, keeping his hands on the covered crate, trying to soothe the still-anxious bird into a few more minutes of silence.

"I've got to get out of here before this thing erupts again," he whispered. "But I just wanted to make sure we're still on for tonight."

Mark raised his eyebrows, and Eduardo nodded, smiling. The night before, he'd met a girl at a Phoenix cocktail hour. Her name was Angie, she was cute and slim and Asian, and she had a friend. Eduardo had convinced her to bring the friend along, and now the four of them were going to meet up for a drink at Grafton Street Grille. A month ago, such a thing would have been almost unthinkable.

"What's her name again?" Mark asked. "The friend, I mean?"

"Monica."

"And she's hot?"

The truth was, Eduardo had no idea if Monica was hot or not. He'd never seen the girl. But in his mind, neither one of them had the right to be so choosy. Up until now, the ladies hadn't exactly been knocking down the doors to get to them. Now that Eduardo was almost in the Phoenix, he was starting to have access to women—and he was determined to bring his friend along with him. He couldn't yet get Mark into the Phoenix himself—but he could certainly introduce him to a girl or two.

Mark shrugged, and Eduardo gently lifted the crate and rose to his feet. As he started down the row toward the aisle, he cast a quick glance back at Mark's outfit—the customary Adidas flip-flops, jeans, and the fleece hoody. Then Eduardo straightened his own tie, brushed chicken feathers off the lapels of his dark blue blazer. The tie and blazer were almost a uniform for him; on days he had meetings for the Investment Association, he even wore a suit.

"Just be there at eight," he called back to Mark as he exited the row. "And, Mark . . ."

"Yeah?"

"Try and wear something nice, for a change."

CHAPTER 5 | THE LAST WEEK OF OCTOBER 2003

Behind every great fortune, there lies a great crime.

If Balzac had somehow risen from the dead to witness Mark Zuckerberg storm into his Kirkland dorm room that monumental evening during the last week of October 2003, he might have amended his famous words; because that historical moment, one that inarguably led to one of the greatest fortunes in modern history, did not begin with a crime so much as a college prank.

If the newly revived Balzac had been there in that spartan, claustrophobic dorm, he might have seen Mark head straight for his computer; there would have been no question that the kid was angry, and that he had with him a number of Beck's beers. As usual, he was probably wearing his Adidas flip-flops and a hoody sweatshirt. It was well known that he pretty much hated any shoes that weren't flip-flops, and one day he was determined to be in a position where those were the only shoes he'd ever have to wear.

Maybe Mark took a deep swig of the beer, let the bitter

taste bite at the back of his throat, as he tapped his fingers against the laptop keyboard, gently summoning the thing awake.

Since high school, it could be observed, his thoughts had always seemed clearer when he let them come out through his hands. To an outside observer, the relationship he had with his computer seemed much smoother than any relationship he'd ever had with anyone in the outside world. He never seemed happier than when he was looking through his own reflection into that glassy screen. Maybe, deep down, it had something to do with control; with the computer, Mark was always in control. Or maybe it was more than that, an almost symbiosis that had grown out of years and years of practice. The way Mark's fingers touched those keys: this was where he belonged. Sometimes, it probably felt like this was the *only* place he belonged.

That evening, at a little after eight P.M., he stared into the brightly lit screen, his fingers finding the right keys, opening up a fresh blog page—something that had most likely been percolating in the back of his mind for a few days. The frustration—likely the result of the evening he had just had—was, it seemed, the final impetus to move further along with the idea, turn the kernel into corn. He started with a title:

Harvard Face Mash/The Process.

He might have looked at the words for a few minutes, wondering if he was really going to go through with this. He might have taken another drink from his beer, and hunched forward over the keys:

8:13 pm: *** is a bitch. I need to think of something to make to take my mind off her. I need to think of something to occupy my mind. Easy enough now I just need an idea.**

Maybe somewhere inside of Mark's thoughts, he knew that blaming it all on a girl who had rejected him wasn't exactly fair. How were this one girl's actions different from the way most girls had treated Mark throughout high school and college? Even Eduardo, geek that he was, had better luck with girls than Mark Zuckerberg did. And now that Eduardo was getting into the Phoenix—well, tonight Mark was going to do something about his situation. He was going to create something that would give him back some of that control, show all of them what he could do.

Perhaps he took another drink, then turned his attention toward the desktop computer next to his laptop. He hit a few keys, and the desktop's screen whirred to life. He quickly opened up his Internet connection, linking himself to the school's network. A few more clicks of the keys, and he was ready.

He turned back to the laptop, and went back to work on the blog:

> **9:48 pm: I'm a little intoxicated, not gonna lie. So what if it's not even 10pm and it's a Tuesday night? What? The Kirkland facebook is open on my computer desktop and some of these people have pretty horrendous facebook pics.**

Maybe he grinned as he scanned through the pictures that were now spread across the screen of his desktop. Certainly, he recognized some of the guys, and even a few of the girls—but most of them were probably strangers to him, even though he'd passed them in the dining hall or on his way to his classes. He was probably a complete stranger to them, too; some of the girls, for sure, had gone out of their way to ignore him.

> **I almost want to put some of these faces next to pictures of farm animals and have people vote on which is more attractive.**

At some point during this process, Mark began to exchange ideas with his friends who had gotten home from dinner, classes, drinks—most of the communication coming, as it usually did, via e-mail. Nobody in his circle used the phone much anymore; it was all e-mail. Other than Eduardo, they were all almost as infatuated with their computers as Mark was. He turned back to the blog:

> **It's not such a great idea and probably not even funny, but Billy comes up with the idea of comparing two people from the facebook, and only sometimes putting a farm animal in there. Good call Mr. Olson! I think he's onto something.**

Yes, to a kid like Mark it must have indeed seemed a great idea. The Kirkland housing facebook—all of the school's facebooks, as their databases of student photos were known—was such a stagnant thing, compiled entirely in alphabetical order by the university.

The percolations that must have gripped Mark's imagination for a few days were now forming into something real—an idea for a Web site. To Mark, it's likely that the cool thing was the math that was going to go into it—the computer science of the task, the code at the heart of the Web-site idea. It wasn't just a matter of writing a program, it was also creating the correct algorithm. There was some complexity to it that his friends would surely appreciate—even if the larger campus of bimbos and Neanderthals never understood.

> **11:09 pm: Yea, it's on. I'm not exactly sure how the farm animals are going to fit into this whole thing (you can't really ever be sure with farm animals . . .), but I like the idea of comparing two people together. It gives the whole thing a very Turing fell, since people's ratings of the pictures will be more implicit than, say, choosing a number to represent**

each person's hotness like they do on hotornot.com. The other thing we're going to need is a lot of pictures. Unfortunately, Harvard doesn't keep a public centralized facebook so I'm going to have to get all the images from the individual houses that people are in. And that means no freshman pictures . . . drats.

Maybe, at this point, he knew that he was about to cross a line—but then, he'd never been very good at staying within the lines. That was Eduardo's game, wearing a jacket and tie, joining that Final Club, playing along with everyone else in the sandbox. From Mark's history, it was obvious that he didn't like the sandbox. He seemed the type who wanted to kick out all the sand.

12:58 am: Let the hacking begin. First on the list is Kirkland. They keep everything open and allow indexes in their Apache configuration, so a little wget magic is all that's necessary to download the entire Kirkland facebook. Child's play.

It really was that simple—for Mark. Most likely, in a matter of minutes, he had all the pictures from the Kirkland facebook downloaded off of the university's servers and into his laptop. Sure, in a sense it was stealing—he didn't have the legal rights to those pictures, and the university certainly didn't put them up there for someone to download them. But then, if information was getable, didn't Mark have the right to get it? What sort of evil authority could decide that he wasn't allowed access to something he so easily could access?

1:03 am: Next on the list is Eliot. They're also open, but with no indexes in Apache. I can run an empty search and it returns all of the images in

the database in a single page. Then I can save the page and Mozilla will save all the images for me. Excellent. Moving right along . . .

He was now deep in hacker's paradise. Breaking into Harvard's computer system really was child's play to him. He was smarter than anyone Harvard had employed to make the system, he was smarter than the administration, and he was certainly smarter than the security systems Harvard had put into place. Really, he was teaching them a lesson—showing them the flaws in their system. He was doing a good deed, though it was pretty likely that they wouldn't have seen it that way. But hey, Mark was documenting what he was doing right there in his blog. And when he built the Web site, he was going to put that blog right there on the site, for everyone to see. Maybe crazy, a little, but that was going to be the icing on the cake.

1:06 am: Lowell has some security. They require a username/password combo to access the facebook. I'm going to go ahead and say that they don't have access to the main fas user database, so they have no way of knowing what people's passwords are, and the house isn't exactly going to ask students for their fas passwords, so it's got to be something else. Maybe there's a single username/password combo that all of Lowell knows. That seems a little hard to manage since it would be impossible for the webmaster to tell Lowell residents how to figure out the username and password without giving them away completely. And you do want people to know what kind of authentication is necessary, so it's probably not that either. So what does each student have that can be used for authentication that the house webmaster has access to? Student ID's anyone? Suspicions affirmed—time to get myself a matching name and student ID combo for Lowell and I'm in. But there are more problems. The pictures are separated into a bunch of different

pages, and I'm way too lazy to go through all of them and save each one. Writing a perl script to take care of that seems like the right answer. Indeed.

It was hacking at its most fundamental—like a cryptographer working out of some cave to defeat the Nazis' code. By now Mark's computer was filling up with pictures; pretty soon he'd have half the house photo database in his hands. Every girl on campus—except the freshmen—under his control, in his laptop, little electronic bytes and bits that represented all those pretty and not so pretty faces, blondes and brunettes and redheads, big-breasted and small, tall and short, all of them, every girl. This was going to be fantastic.

1:31 am: Adams has no security, but limits the number of results to 20 a page. All I need to do is break out the same script I just used on Lowell and we're set.

House by house, name by alphabetical name. He was collecting them all.

1:42 am: Quincy has no online facebook. What a sham. Nothing I can do about that. 1:43 am: Dunster is intense. Not only is there no public directory, but there's no directory at all. You have to do searches, and if your search returns more than 20 matches, nothing gets returned. And once you do get results, they don't link directly to the images; they link to a php that redirects or something. Weird. This may be difficult. I'll come back later.

The houses he couldn't get through right away, he'd most likely figure out later. There was no wall that he couldn't climb. Harvard was the premier university in the world, but it was no match for Mark Zuckerberg, for his computer.

1:52 am: Leverett is a little better. They still make you search, but you can do an empty search and get links to pages with every student's picture. It's slightly obnoxious that they only let you view one picture at a time, and there's no way I'm going to go to 500 pages to download pics one at a time, so it's definitely necessary to break out emacs and modify that perl script. This time it's going to look at the directory and figure out what pages it needs to go to by finding links with regexes. Then it'll just go to all of the pages it found links to and jack the images from them. It's taking a few tries to compile the script . . . another Beck's is in order.

Mark was most likely wide-awake now, deep into the process. He didn't care what time it was, or how late it got. To guys like Mark, time was another weapon of the establishment, like alphabetical order. The great engineers, hackers—they didn't function under the same time constraints as everyone else.

2:08 am: Mather is basically the same as Leverett, except they break their directory down into classes. There aren't any freshmen in their facebook . . . how weak.

And on and on he went, into the night. By four A.M., it seemed as though he had gone as far as he could go—downloaded thousands of photos from the houses' databases. It was likely that there were a few houses that weren't accessible online from his James Bond–like lair in Kirkland house—you probably needed an IP address from within these houses to get at them. But it was also likely that Mark knew how to do that—it would just take a little legwork. In a few days, he could have everything he needed.

Once he had all the data, he'd just have to write the algorithms. Complex mathematical programs to make the Web site work. Then the program itself. It would take a day, maybe two at the most.

He was going to call the site Facemash.com. And it was going to be beautiful:

> **Perhaps Harvard will squelch it for legal reasons without realizing its value as a venture that could possibly be expanded to other schools (maybe even ones with good-looking people). But one thing is certain, and it's that I'm a jerk for making this site. Oh well. Someone had to do it eventually . . .**

Maybe grinning as he downed the rest of his Beck's, he spelled out the introduction that would greet everyone who went to the site when he finally launched it:

> **Were we let in for our looks? No. Will we be judged on them? Yes.**

Yes, it was going to be fucking beautiful.

CHAPTER 6 | LATER THAT EVENING

If you were to ask the right computer hacker what might have happened next, that frigid fall night in Cambridge, the answer seems fairly clear. Based on the blog he created, documenting his thought process as he created Facemash, one can surmise what might have followed. Maybe there are other explanations, but we know there were certain houses Mark was having trouble hacking into. He might have gotten what he needed in other ways, we certainly don't know for sure every detail; but we can imagine how it might have gone down:

A Harvard residence house. The middle of the night. A kid who knew a lot about computer security and how to get around it. A kid who lived outside the great, churning hormonal world of college life. Maybe a kid who wanted inside. Or maybe a kid who just liked to prove what he could do, how much smarter he was than everybody else.

Imagine the kid, crouching in the dark. Down real low, hands and knees low, curled up in a deep crouch behind a velvet couch. The carpet beneath his fingers and flip-flops is plush and crimson, but most of the rest of the room is just shadows, a twenty-by-twenty cavern of shapes and silhouettes.

Maybe the kid isn't alone—maybe two of the shapes are people, a guy and a girl, positioned by the far wall, right between the windows that looked out onto the house's courtyard. From his position behind the couch, the kid wouldn't have been able to tell if they were sophomores, juniors, or seniors. But he would have known they are trespassing—just like he is. The third-floor parlor isn't exactly off-limits, but normally you needed a key to get inside. The kid didn't have a key, he'd just timed it perfectly—waiting outside the door on the third-floor landing for the janitor to finish with the carpet and the windows, and then, just at the right moment, as the man packed up his gear and walked out—lunging inside, leaving a textbook levered in the door frame.

The guy and the girl, on the other hand, had just gotten lucky. They'd probably noticed the door propped open, and curiosity had driven them forward. As we imagine it, the kid had ducked behind the couch just in time. Not that the couple is going to notice him—they have other things on their minds.

At the moment, the guy has the girl back against the wall, her leather jacket open and her sweatshirt up all the way past her collarbone. The guy's hands are moving up her flat, naked stomach, and she arches her back, his lips touching the side of her throat. She seems about to give in to him, right then and there—but thankfully, something makes her change her mind. She lets him go a second longer, then pushes him away and laughs.

Then she grabs his hand and drags him back across the room, toward the door. They pass right by the couch—but neither of them looks in the kid's direction. By the time the girl reaches the doorway and pushes the door open, the guy has his arm around her waist, and he half carries her out into the hall. The door swings back against the textbook—and for a brief second, the kid thinks the textbook is going to slip out and he'll be locked inside all night. Thankfully, the book holds. And finally the kid is alone, with the shadows and the silhouettes.

We imagine him slipping out from behind the couch and continuing what he'd been doing before he'd been interrupted. He begins prowling around the perimeter of the room, his knees slightly bent as he scans the dark walls, especially the area right down by the molding. It takes another few minutes to find what he's looking for—and then he grins, reaching for the small backpack that is slung over his left shoulder.

He gets down on his knees as he opens the backpack. His fingers find the little Sony laptop inside, and he yanks the device free. An Ethernet cable is already attached to the Sony, swinging free and pendulant as he powers the machine up. With expert ease, he catches the end of the cable—and jams it into the port in the wall, a few inches above the plaster molding.

With quick flicks of his fingers against the computer's keyboard, he engages the program he'd written just a few hours ago and watches as the laptop screen blinks up at him; with him, we can almost imagine the tiny packets of electrical information siphoning upward through the cable, tiny pulses of pure energy culled from the electronic soul of the building itself.

The seconds tick by as the laptop whirs in near silent gluttony, and every now and then the kid glances behind himself, making sure the room is still empty. His heart is no doubt pounding, and we can imagine the tiny rivulets of sweat trickling down the small of his back. We don't think this is the first time he's done something like this, but the adrenaline high is always the same; it must feel like James Bond kind of shit. Somewhere in the back of the kid's mind he must know that what he is doing is probably illegal—certainly against school rules. But it isn't exactly Murder One. As hacking goes, it isn't even shoplifting.

He isn't stealing money from a bank, or hacking into some Defense Department Web site. He isn't fucking with some power company's grid or even tracking some ex-girlfriend's e-mail. Considering what a highly trained hacker such as himself is truly capable of, he is hardly doing anything at all.

Just taking a few pictures off of a house database, that's all. Well, maybe not just a few pictures—*all of them.* And maybe it is a private database, one that you are supposed to have a password to access—and an IP address from this particular building along with that password to comb through—okay, it isn't totally innocent. But it isn't a capital crime. And in the kid's mind, it is certainly for the greater good.

A few more minutes and he'll be done. *The greater good.* Freedom of information and all that crap—to him, we believe, it is part of a true moral code. Kind of an extension of the hacker's creed—if there's a wall, you find a way to knock it down or crawl over it. If there's a fence, you cut your way through. The people who built the walls, the "establishment"—they are the bad guys. The kid is the good guy, fighting the good fight.

Information is *meant* to be shared.

Pictures are *meant* to be looked at.

A minute later, a tiny electronic beep emerges from the laptop, signaling that the job is finished. The kid pops the Ethernet cable out of the wall and jams the laptop back into his backpack. This house down, maybe two houses to go. We almost hear the James Bond theme running through the kid's head. He slings the backpack over his left shoulder and hurries toward the door. He retrieves the textbook, slips out of the parlor, and lets the door click shut behind him.

We can imagine him noticing, as he goes, that the girl's floral perfume still hangs, seductively, in the air.

CHAPTER 7 | WHAT HAPPENS NEXT

It wasn't until about seventy-two hours later that Mark found out exactly what he'd done. His drunken evening had most assuredly long since subsided; but he'd carried through with what he'd started, even while he'd gone on about his life, going to his computer science courses, studying for his Cores, hanging with Eduardo and his buddies in the dining hall. Later on, he'd tell reporters from the college newspaper that he hadn't even thought that much about Facemash, other than that it was a task to be completed, a mathematical and computing problem to be solved. And when he'd done that—perfectly, wonderfully, beautifully—finishing up just a couple of hours earlier, he'd e-mailed it to a few of his buddies to see what they thought. To get opinions, feedback, maybe even a few accolades. Then he'd headed out of his room to a meeting for one of his classes, which had lasted a good deal longer than he'd expected.

By the time he'd gotten back to his dorm in Kirkland, all he had intended to do was drop off his backpack, check his e-mails, and head down to the dining hall. But when he entered his bedroom, his attention immediately slid to the laptop that was still open on his desk.

To his surprise, the screen was frozen.

And then it dawned on him. The laptop was frozen because it was acting as a server for Facemash.com. But that didn't make any sense, unless—

"Holy shit."

Before he left for the meeting, he had e-mailed the link to Facemash.com to a handful of friends. But obviously, some of them had forwarded it along to their friends. Somewhere along the way, it had picked up steam of its own. From the program trail, it looked like it had been forwarded to a dozen different e-mail lists—including some lists run by student groups on campus. Someone had sent it to everyone involved with the Institute of Politics, an organization with over a hundred members. Someone else had forwarded it to Fuerza Latina—the Latina women's issues organization. And someone from there had forwarded it to the Association of Black Women at Harvard. It had also gone to the *Crimson,* and had been linked on some of the house bulletin boards.

Facemash was everywhere. A Web site where you compared two pictures of undergraduate girls, voted on which one was hotter—then watched as some complex algorithms calculated who were the hottest chicks on campus—had gone viral throughout the campus.

In under two hours, the site had already logged twenty-two thousand votes. Four hundred kids had gone onto the site in the past thirty minutes.

Shit. This wasn't good. The link wasn't supposed to go out like that. He'd later explain that he wanted to get some opinions, maybe tweak the thing a bit. He'd wanted to figure out what the legalities were of downloading all those pictures. Maybe he'd never have launched it at all. But now it was too late. The thing about the Internet was, it wasn't pencil, it was pen.

You put something out there, you couldn't erase it.

Facemash was out there.

Mark lunged forward, hitting keys on the desktop, using passwords to get inside the program he'd written. In a matter of minutes, he killed the damn thing, shutting it down. He watched as his laptop screen finally went blank. Then he dropped down into his chair, his fingers trembling.

He had a feeling that he was in big trouble.

CHAPTER 8 | THE QUAD

From the outside, the four-story Hilles Building looked more like a crash-landed space station than a university library; jutting pillars of cement and stone, shiny facades of glass and steel. Like the rest of the Quad, the Quad Library was one of the newer buildings on campus; because it was tucked so far away from the Yard and its aging, ivy-covered legacy, the architects had probably figured they could get away with just about anything. Even a futuristic monstrosity that seemed more appropriate for the MIT campus down the street.

At the moment, Tyler was entombed in a back corner on the third floor of the spaceship, his six-foot-five frame jammed into a chair-desk combination that seemed almost as much torture device as a piece of Art Deco furniture. He'd chosen the chair-desk monstrosity specifically because it was uncomfortable; it was barely seven in the morning on a Monday, and after the workout he'd just had, it was going to take extraordinary measures just to keep himself conscious.

There was a massive economics textbook open on the desk in front of him, next to one of the bright red plastic trays from the nearby Pforzheimer House dining hall. A half-eaten bologna sandwich was on the tray, partially wrapped in a napkin. Even though Tyler and Cameron had just finished breakfast not a half hour ago, Tyler was still starving; the textbook was the reason he was in the library, with less than an hour to go before his Econ 115 lecture—but the bologna sandwich was the only thing keeping him awake. The missing half of the sandwich was still in his mouth, and he was so busy chewing that he didn't even hear Divya approach from behind.

Seemingly out of nowhere, Divya reached over his shoulder and slammed a copy of the *Crimson* down onto the plastic tray—sending what was left of the bologna sandwich spinning off toward the floor.

"I'm not going to find us a computer programmer in the *Crimson*?" Divya half shouted, by way of a greeting. Tyler glared up at him, a masticated chunk of meat hanging from his mouth.

"What the fuck, man?"

"Sorry about the sandwich. But look at the headline."

Tyler grabbed the newspaper and shook ketchup off of its back page. He glared at Divya again, then he looked where his Indian friend was pointing. Tyler's eyebrows rose as he shifted from the headline to the article, quickly skimming the first few paragraphs.

"Okay. This is pretty cool," he acknowledged.

Divya nodded, grinning. Tyler leaned back in his chair and stretched his neck so that he could see around the corner. He could just make out Cameron's long legs stretched out from beneath an identical chair-desk combination, not ten feet away.

"Cameron, wake up and get your ass over here!"

A few nearby students looked up, saw that it was Tyler—then went back to their work. It took Cameron a few moments to disentangle himself from the chair-desk, but eventually he plodded over and took position next to Divya. Cameron's hair was standing up in the back, and his eyes were bleary and red. The wind on the river had been pretty fierce that morning, and crew practice had been particularly brutal. But Tyler no longer felt anywhere near as tired as his brother looked, not after seeing what Divya had shown him.

Tyler handed Cameron the paper. Cameron glanced at the article, nodding.

"Yeah, I heard some of the guys at the Porc talking about this last night. Sam Kensington was pretty pissed off, because his girlfriend Jenny Taylor got ranked number three by the site, and her roommate Kelly was number two—"

"And her other roommate Ginny was ranked number one," Divya interrupted. "Not that anyone was surprised."

Tyler had to smile. Jenny, Kelly, and Ginny were arguably the hottest three sophomore girls on campus. They'd been freshman roommates as well, put together supposedly at random. Except nobody on campus really believed it was random—especially since someone figured out that the last five digits of their freshman dorm-room phone number turned out to be "3-FUCK." The Harvard housing office was notorious for bizarre little pranks like that. Putting kids with similar names in the same rooms. Tyler's freshman year, there was a Burger and Fries, and at least two Blacks and Whites. And then there was Jenny, Kelly, and Ginny, the three hottest blondes on campus, in a room with the phone number 3-FUCK. Someone probably needed to get fired.

But the housing office wasn't the subject of the *Crimson* article. The three blondes had been ranked by a Web site—according to the *Crimson,* it had been called Facemash, a sort of "hot-or-not" clone where students were able to rate girls based on their pictures—and it had caused quite a stir on campus.

"It got shut down pretty fast," Divya continued, pointing to the *Crimson.* "Says here that the kid who made it shut it down himself. When he created the site, he didn't even realize people were going to get mad. Even though on his blog, he talked about comparing girls to farm animals."

Tyler leaned back in his chair.

"Who got mad?"

"Well, girls. Lots of them. The feminist groups on campus sent dozens of letters. And then the university—so many people were on the site at the same time, it clogged up the university's bandwidth. Professors couldn't even get into their e-mail accounts. It was a righteous mess."

Tyler whistled low.

"Wow."

"Yeah, wow. It got like twenty thousand hits in twenty minutes. Now the kid who created it is in a lot of trouble. Seems he stole all the pictures off the houses' databases. Hacked in and just downloaded 'em all. Him and a few of his friends are gonna get ad-boarded."

Tyler knew all about the ad board—the administration's disciplinary organization, usually made up of deans and student advisers, sometimes even university lawyers and the higher administrators themselves. Tyler had a friend in the Porc who'd been accused of cheating on a history exam. The kid had had to go up in front of two deans and a senior tutor. The ad board had a lot of power—it could suspend you, even call for expulsion. Though in this case, Tyler doubted the punishment would be that severe.

The kid who made Facemash would probably end up getting probation. His reputation was a little fucked, however. Certainly the girls on campus weren't going to be thinking too highly of him. Although, from the sound of it, the kid wasn't exactly a Casanova.

Comparing farm animals to girls? That wasn't the sort of thing you came up with when you were getting laid regularly.

"Says here it's not his first program," Cameron said, leafing through the article. "He wrote that Course Match thing. You remember, Tyler, that online schedule, to pick your classes. And in high school, he was supposedly some sort of megahacker."

Tyler felt the energy rising inside of him. He liked everything he was hearing. This kid had fucked things up with his Web site—but he was obviously a brilliant programmer, and definitely a freethinker. Maybe he was exactly what they were looking for.

"We should talk to this kid."

Divya nodded.

"I already called Victor. He says the kid is in some computer science classes with him. He warned me that he's a little weird, though."

"Weird how?" Cameron asked.

"You know, like kind of socially autistic."

Tyler looked at Cameron. They knew exactly what Divya meant. *Autistic* wasn't the right word; *socially awkward* probably covered it. There were dozens of kids like that all over Harvard. To get into Harvard, you had to either be incredibly well rounded—like a straight-A student who was also the captain of a varsity sport. Or you had to be really, really, really good at one thing—maybe better than anyone else in the world. Like a virtuoso violinist, or an award-winning poet.

Tyler liked to think that he and his brother were well rounded—but he wasn't going to fool himself, he also knew that they were really, really, really good at crew.

This kid was obviously really, really, really good at computers—because it sounded like he sure as hell wasn't the captain of any varsity sport.

"What's the kid's name?" Tyler asked, his mind already whirring ahead.

"Mark Zuckerberg," Divya answered.

"Go send him an e-mail," Tyler decided, tapping at the *Crimson.* "Let's see if this Zuckerberg kid wants to be a part of history."

CHAPTER 9 | THE CONNECTION

From the steps of Widener Library, in the bright light of eleven A.M., Harvard Yard looked pretty much as it had for the past three hundred years. Little tree-lined paths meandering between patches of meticulously shorn grass. Ancient, brick-and-stone buildings covered in ivy, complicated twists of green that curled like veins across aging architectural skin. From Eduardo's vantage on the top stone step, he could just make out the peak of Memorial Church in the distance, but nothing beyond, not the space-age science center or boxy freshman dorm Canaday, none of the newer buildings that marred the austere continuum of the historical-minded campus. There was weight in that view, centuries of moments like these—though Eduardo had a feeling, in all those hundreds of years, no student had lived through precisely the sort of bizarre torture that the kid sitting next to him had just endured.

He looked over at Mark, who was sitting cross-legged next to him on the step, partially enveloped in a shadow cast by one of the vast pillars that held up the roof of the great stone library. Mark was wearing a suit and tie, and he appeared uncomfortable, as usual—but at the moment, Eduardo was pretty sure his friend's discomfort was due only partially to his clothes.

"That was unpleasant," Eduardo commented, turning his attention back to the Yard.

He watched a pair of pretty freshmen hurrying down one of the paths. The girls were wearing matching Crimson scarves, and one had her hair up in a bun, showing off a porcelain stretch of neck.

"Kind of like a colonoscopy," Mark responded.

He was watching the girls' progress across the yard as well. Maybe he was thinking the same thing that Eduardo was—that those girls had probably heard of Facemash, maybe read about it in the *Crimson* or seen something posted on one of the online campus bulletin boards. Maybe the girls were even aware that just an hour before, Mark had been forced to sit down in front of the ad board and explain himself, that he'd been propped up in front of no fewer than three deans, not to mention a pair of computer security experts, and made to apologize, again and again, for the mess he had inadvertently caused.

The funny thing—although the deans hadn't exactly seen the humor in it—was that Mark hadn't seemed to really understand why anyone was so upset in the first place. Yes, he'd hacked into the university's computers, and he'd downloaded pictures. He knew that was wrong, and he'd certainly apologized for it. But he was truly confused by the anger that had been directed toward him by the various female groups on campus—and not just the groups, but by the girls themselves, many of whom had sent e-mails, letters, and sometimes boyfriends to get the message across. In the dining hall, in the classrooms, even in the library stacks, wherever they ran across Mark's path.

During the ad-board meeting, he'd readily admitted his guilt in terms of the hacking—but he'd also pointed out that his actions had illuminated some serious security flaws in the university's computer system. His stunt had a silver lining, he'd argued, and he'd readily volunteered to help the houses fix up the flaws in their systems.

Also, Mark had gamely pointed out that he'd shut the site down himself, when he'd realized it had gone viral. He'd never had any intention of launching Facemash across the campus—it was sort of a beta test gone wild. A stunt, and he hadn't meant to do anything malicious with the Web site.

Frankly, Mark's social awkwardness—and his confusion over the response to Facemash—had been his greatest defense. The gathered deans had looked at him and listened to his stilted affectation, and they had realized that Mark really wasn't a bad kid—he just didn't think the same way other kids did. He hadn't realized that girls were going to get mad because guys were voting on their appearance—hell, Mark and Eduardo and probably every other college guy in the world had been ranking female classmates in terms of hotness since the dawn of structured education. Eduardo was pretty sure that someday, some paleontologist would find a cave drawing ranking Neanderthal girls—it was simply human nature to make that kind of list.

To an outside observer, it seemed that Mark hadn't realized that the sort of things that went on in his mind, the sort of conversations you had with your fellow geek friends in the privacy of your geek lairs—they didn't play well out in the general public. You suggest putting pictures of girls up against farm animals, and you're going to piss people off.

Mark had certainly pissed a lot of people off. But the deans, in their good graces, had decided not to suspend or expel him over Facemash. They'd given Mark a form of probation—really, they'd simply told him to stay out of trouble for the next two years, or else. They hadn't clearly defined what "or else" meant, but in any event, it was a good, solid slap on the wrist.

Mark had survived the incident without much damage to his academic standing. His reputation on campus, however, hadn't gotten off quite as easily. If he'd had trouble getting girls before, he was going to have a hell of a time with them now.

Then again, people knew the name Mark Zuckerberg. The *Crimson* article had made sure of that. The paper had even followed up the initial article about the Facemash debacle with an editorial about Facemash's popularity, and how the very number of hits the site had garnered showed that there was interest in a sort of online picture-sharing community—though maybe not one with such a negative bent. Mark had certainly stirred up the pot—that was something, wasn't it?

When the two freshman girls strolled out of view, Mark reached into his back pocket and pulled out a piece of folded paper, then turned to Eduardo.

"I want to show you something. What do you think of this?"

He handed the paper over, and Eduardo unfolded it; it was an e-mail, printed off of Mark's computer:

> **Hey Mark, I got your email from my friend. In any case, me and my team need a web developer with php, sql, and hopefully java sills. We're very deep into developing a site, which we'd like you to be a part of and a site which we know will make some waves on campus. Please call my cell or write me an email letting me know when you'd be free to chat on the phone and meet with our current developer. This should be a really rewarding experience, especially if you have an entrepreneurial personality. We'll let you know the details when you respond. Cheers.**

The e-mail was signed by someone named Divya Narendra, and had been cc'd to someone named Tyler Winklevoss. Eduardo read through the e-mail twice, digesting the request. It sounded like these kids were working on some sort of secret Web site—probably they had read about Mark in the *Crimson,* had seen Facemash, and were thinking he could help them with whatever it was they were building. They certainly didn't know Mark—they were responding to his reputation, his sudden notoriety.

"You know these guys?" Mark asked.

"I don't know Divya, but I know who the Winklevoss twins are. They're seniors, I think they live in the Quad. They row crew."

Mark nodded. Of course, he knew the Winklevoss twins, too. Not personally, of course, but you couldn't avoid having noticed them at some point. Six-foot-five identical twins were kind of hard to miss. But neither Eduardo nor Mark had ever exchanged a word with the two jocks; they weren't exactly wandering around in the same circles. Tyler and Cameron were Porc guys. They were athletes, and they hung out with athletes.

"Are you going to talk to them?"

"Why not?"

Eduardo shrugged. He glanced at the e-mail again. To tell the truth, he didn't have a great feeling about it. He didn't know the Winklevoss twins, or Divya, but he knew Mark, and he couldn't imagine Mark getting along well with kids like that. It took a certain sort of "understanding" to get along with Mark in the long run. And guys like the Winklevosses, well, they didn't understand geeks like Eduardo and Mark.

Sure, Eduardo was making great ground now that he was hanging out at the Phoenix, working his way through the initiation process. In a week or so, he was pretty sure that process would end—

and he'd become a full-fledged Final Club member. But there was a vast difference between being a member of the Phoenix and being a member of the Porcellian. The Phoenix was about learning how to talk to girls, drink heavily, and hopefully get laid. The Porcellian was about learning how to rule the world.

"I'd say fuck 'em," Eduardo responded. "You don't need them."

Mark took the e-mail back and shoved it into his pocket. Then he picked at his shoelaces, loosening his shoes.

"I don't know," he said, and Eduardo could tell that he'd already made up his mind. Maybe the idea of hanging out with guys like the Winklevoss twins appealed to something inside Mark, or maybe it was just another lark, like Facemash—something that seemed like it could be amusing.

Or, as Mark would put it, as he always put it:

"It might be interesting."

CHAPTER 10 | NOVEMBER 25, 2003

"Oh, shit. Lock up your girlfriends, boys. Look who's coming to dinner."

Tyler and Cameron were halfway through the Kirkland dining hall, moving between the tables at a near jog, when it happened. Tyler saw the bull-shaped senior coming toward them, hands outstretched in a low, faux tackle, a sloppy grin above those wide, saggy jowls—and he just had to laugh. The very idea that they could get through the meeting at the river house without being noticed was foolish; both he and Cameron had a lot of friends in Kirkland, including a few members of the Porc, and a handful of crew teammates. Davis Mulroney wasn't either; but he was hard to avoid, considering that he must have weighed close to three hundred pounds, played center on the varsity football team—and now he was coming right at them.

Tyler feinted left, but he was too slow, and Davis got him in a waist-high bear hug, lifting his feet off the floor for a full count of five. After letting Tyler down, he shook both brothers' hands, then cocked a bushy eyebrow at them.

"Slumming on the river? What brings you boys down from the Quad?"

Tyler glanced at Cameron. They'd both agreed that it was better, for now, to keep their meeting with the computer kid under wraps. It wasn't like their Web site was a complete secret; their friends knew about it, and so did a few of their brothers at the Porc. But this Zuckerberg kid was kind of a flash point on campus at the moment, and they certainly weren't ready for any *Crimson*-level announcements.

Hell, they hadn't even met the kid yet—but they did know he was very interested in their site and wanted to be a part of what they were building. Both Divya and Victor Gua had traded a bunch of e-mails with the kid, and according to them, Zuckerberg had seemed really interested. His exact words in one of his recent e-mails made it sound like he was certainly worth the trip to the river house:

> **I'm down to chat, but I need to deal with the aftermath of facemash—so maybe tomorrow? I'm definitely interested in hearing about your project.**

But a dinner meeting at Kirkland wasn't the same as a full partnership, and Tyler didn't need the whole campus knowing that he and his brother were working with the Facemash kid before it was actually true. Still, it was foolish to think that he and his brother could march into Kirkland without running into a handful of friends. Davis's girlfriend was roommates with one of Cameron's exes; and anyway, football and crew had similar workout schedules, so they were always running into each other.

"We heard it was sloppy-joe night," Tyler responded. "We're always up for a good sloppy joe."

Davis laughed. He gestured toward a table near the windows,

which was filled with rather large-looking guys in matching Harvard athletic sweatshirts.

"Why don't you join us? We're gonna grab a drink afterward at the Spi, maybe head down to Grafton. My buddy has some chicks coming in on the Fuck Truck from Wellesley. Should be a good time."

Tyler rolled his eyes. The "Fuck Truck" was a Harvard institution—a vanlike bus that traveled between the Harvard campus and a half dozen of the nearby all-girl schools—as well as a few of the more liberal-minded coed party campuses—shuttling kids back and forth, most often on weekends. All socially knowledgeable Harvard grads had been on the Fuck Truck at least once in their college career; Tyler could close his eyes and still remember the wonderfully thick scent of alcohol and perfume that seemed to permeate the bus's vinyl seats. But tonight, he wasn't interested in the Fuck Truck, or its contents.

"Sorry, man, can't tonight, but maybe a rain check."

He gave the big kid a pat on the shoulder, waved at the table of jocks, then kept on moving through the dining hall. As he went, he couldn't help thinking that in some ways, the Fuck Truck was analogous to the project he and his brother were working on; the Harvard Connection would have features that could be described as an electronic Fuck Truck—a superslick connection between guys and girls, but instead of a long ride in the back of a bus, you'd just have to click a key on your laptop. One-stop shopping, as it were, for that coed of your dreams.

Cameron tapped his arm and pointed toward a table at the very back of the rectangular hall. In the center of the table, a kid was waving at them. The kid was lanky and had a mop of curly brownish blond hair. He was wearing a zippy and cargo shorts, even though it was thirty degrees outside, and his cheeks had a certain ivory pallor to them, as if he hadn't been in the sun in a long time.

There was another kid at the table with him—a short, dark-haired guy with scruff on his chin, maybe the kid's roommate—but that one took off as they approached, leaving Mark by himself. Tyler reached the table first, holding out his hand.

"Tyler Winklevoss. This is my brother, Cameron. Sorry Divya couldn't make it, he had a seminar he couldn't get out of."

Mark's hand felt like a dead fish in his grip. Tyler dropped into a seat across the table from him, and Cameron took the seat to Tyler's right. Mark didn't look like he was going to say anything, so Tyler started right in.

"We're gonna call it Harvard Connection," he began, getting right to the point. Then he launched into a full description of the Web site they were trying to build. He tried to keep it simple, at first—explaining the idea behind an online meeting place where Harvard guys and girls could find each other, share information, connect. That the site would have two sections, one for dating, and one for connecting. Students would be able to post pictures of themselves, put in some personal info, and try to find links with one another. Then he got into the ideology behind the site—the thought that there was an inefficiency in the way people met each other, how there were so many obstacles to finding the perfect person, how the Harvard Connection could bring people together based on their personalities—or whatever they put online—rather than on their proximity.

Although it was hard to read the kid's face, it seemed like Mark got the idea right away. He liked the concept of a Web site to meet girls, and he was certain that the programming wasn't going to be too difficult for him. He asked how far along Victor had gotten with the code, and Cameron suggested that he could see for himself—they would give Mark the necessary passwords to go inside Victor's work, and he could even download the code so he could work on it from his own computer. Cameron guessed they were talking about ten, maybe fifteen hours of programming left to do—no heavy lifting for a guy like Mark. Cameron went into more detail as Tyler leaned back in his chair, watching as the kid listened.

He could see that Mark was getting more and more excited about the idea as his brother talked. The awkwardness in him seemed less apparent the more into the computer stuff they got, and unlike the other computer science types they'd spoken to, Mark seemed to share the energy and vision that Tyler and his brother had brought to the table. Still, Tyler knew that the kid would want to know what was in it for him if he made the site work, so Tyler jumped into it as soon as his brother quieted down.

"If this site is successful, we're all going to make money," he said. "But more than the money, this is going to be very cool for all of us. And we want you to be the centerpiece of it all. This will get you back in the *Crimson*—but this time, the paper is going to be praising you, not trashing you."

The offer was pretty simple, in Tyler's view. They'd be partners in the project, so if it made any money, they'd all do well. But until then, Mark could use the launch of the Web site to rehabilitate his image. And he could be the center of attention—something that computer guys never really got, as they were often shoved into the background—and use the site however he wanted to better his social situation.

Looking at the kid, alone in the back of his dining hall, obviously awkward, as if uncomfortable in his own skin—Tyler knew that it had to be a seductive thought. Get the site going, get a little famous because of it—who knows, maybe it would make this kid a whole different person. Give him a social life, break him out of the geeky mold, get him in with the type of girls you couldn't get hanging out in a computer lab.

Tyler didn't know the kid at all—but who wouldn't respond to an offer like that?

By the time the meeting was over, Tyler knew that the kid was hooked. When they shook hands again, it was less dead fish and more lively engineer—and Tyler headed away from the table thrilled to have finally made contact with someone who seemed to really understand what they were trying to do.

He was so thrilled, in fact, that he decided he and his brother did have time to join the football kids for one drink at the Spi. The Harvard Connection was one step closer to reality, maybe it was time for a little celebrating.

And what could be more fitting a celebration than a visit from the Fuck Truck?

CHAPTER 11 | CAMBRIDGE, 1.

On a good day, the fierce aroma of roasted garlic and Parmesan cheese wafting out of the chrome-and-glass open kitchen would have been titillating, if a little overwhelming. But today was anything but a good day. Eduardo's head was throbbing, and his eyes burned like they'd been dipped in bleach. The aroma was choking him, and he wanted nothing more than to crawl underneath the table in the small booth where he was sitting, curl up into a ball on the floor, and drift off into a coma. Instead, he took deep sips from the glass of ice water he had in front of him and tried to make sense of the blurred words spread across the small menu in his hands.

He didn't blame the restaurant for his physical state; Cambridge, 1. was one of his favorite eateries in Harvard Square, and usually he looked forward to their thick, piled-on pizzas. You could smell Cambridge, 1. from two blocks down Church Street, and there was a good reason every booth in the modern little place was filled, as well as every seat at the small bar that sidled up next to the open kitchen. But at the moment, Eduardo had no interest in pizza. The very thought of food threatened his fragile equilibrium, and he fought the urge to sprint back to his dorm room, cover himself in his blanket, and disappear for the next two days.

He could have gotten away with it, too. It was only a week into January, and he hadn't even started classes yet, after the two-week winter break. In fact, he'd only gotten back onto campus from Miami the day before. After landing at Logan, he'd headed directly over to the Phoenix—really, to decompress after so much family time.

Eduardo had returned to campus needing a mind-cleansing experience—and he'd had no trouble finding one at the Phoenix. He'd also found a few of his fellow new members there, and they'd thrown things right into high gear. It was almost as if they were trying to re create the damage that had been done the night of their initiation into the club—which had occurred just ten days earlier.

Eduardo grinned, even through his pain, as he thought back to that night—truly, one of the craziest of his life. It had started innocuously enough; dressed up in a tuxedo, he and the other initiates had been marched like dapper soldiers all over Harvard Square. Then they had been herded back to the mansion on Mt. Auburn Street and brought into the upper living room of the clubhouse.

The rituals had kicked off with an old-fashioned boat race; the initiates had been divided into two groups, lined up in front of the pool table—and the first kid in each group had been handed a bottle of Jack Daniel's. One of the club members had blown a whistle, and the race had begun. Each initiate had been told to drink as much as he could—then pass the bottle back to the next kid in line.

Sadly, Eduardo's team hadn't won the race—and as punishment, they'd had to reenact the damn thing with an even bigger bottle of vodka.

After that, Eduardo's memory of the night was kind of blurry—

but he did remember being marched out to the river, still wearing his tux. He remembered how fucking cold it was, standing there in his thin little jacket, the December wind whipping through his expensive white shirt. Then he remembered the brothers telling him and the other initiates that they were going to race again—except this time, it was going to be a swimming race. Across the Charles and back.

Eduardo had nearly fainted at the idea. The Charles was notoriously polluted—and even worse, in the middle of December, it was already starting to ice in some places. Trying to swim across sober was terrifying enough—but drunk?

Still, Eduardo hadn't had a choice. The Phoenix meant too much to him to turn back then—so like the other initiates, he'd gone to work on his shoes and socks. Then he'd lined up right at the water's edge, leaned forward—

And, thank God, that's when the brothers had all come out of the darkness, laughing and cheering. There wasn't going to be any swimming that night—just more drinking, more rituals, and congratulations all around. Within a few hours, the initiation was complete, and Eduardo became an official member of the Phoenix.

He was now free to wander the upstairs halls and private rooms of the club, free to get acquainted with the nooks and crannies of the mansion where he'd be spending so much of his social life going forward. To his surprise, last night he'd discovered that there were even bedrooms upstairs in the club—even though nobody actually lived there. He could guess what the bedrooms were for—and the thought had led to many more toasts with his club mates—which had led to the terrifying state he was in now.

So bad, in fact, that he was halfway out of the booth and heading for the door when he finally spotted Mark winding past the crowded bar, his hoody up over his head, a strange, determined glow in his eyes. Eduardo immediately decided he could fight the pain for a few minutes, at least; it wasn't often he'd seen that look in Mark's eyes, and it could only mean something "interesting" was about to go down. Something, at the very least, that would explain why they were meeting in an Italian restaurant instead of in the dining hall, where they usually ate lunch.

Mark slid into the booth across from Eduardo just as Eduardo repositioned himself back behind his ice water and his menu. But from the look on Mark's face, he didn't think they'd be ordering anything soon. Mark seemed to be bursting at the seams.

"I think I've come up with something," he started, and then he launched right into it.

Over the past month—beginning right after the Facemash incident—Mark had been developing an idea. It had really started with Facemash itself—not the Web site per se, but the frenzied interest that Mark had witnessed, firsthand. Simply put, people had reacted to the site—in droves. It wasn't just that Mark had put up pictures of hot girls onto the Internet—there were a million places people could go to see pictures of hot girls—but Facemash had offered up pictures of girls whom the kids at Harvard knew, sometimes personally. The fact that so many people had clicked onto the site, and voted, showed that there was real interest in checking out classmates in an informal, online setting.

Well, Mark wondered, if people wanted to go online and check out their friends—couldn't they build a Web site that offered exactly that? An online community of friends—of pictures, profiles, whatever—that you could click into, visit, browse around. A sort of social network—but one that was exclusive, in that you had to know the people on the site to get into it. Kind of like in the real world—real social circles—but put online, by the people in the social circles themselves.

Unlike Facemash, he wanted to create a Web site where people put their own pictures up—and not just pictures, but also profiles. Where they'd grown up, how old they were, what they were interested in. Maybe the classes they were taking. What they were looking for online—friendship, love interests, whatever. And then he wanted to give people the ability to invite their friends to join. Punch them, in a way, and invite them into your online social circle.

"I'm thinking we keep it simple and call it the Facebook," Mark said, and his eyes were positively on fire.

Eduardo blinked, his hangover suddenly forgotten. Right away he thought it was a pretty amazing idea. It felt big—even though aspects of it certainly sounded familiar. There was a Web site called Friendster that seemed similar, but it was pretty clunky and nobody used it, at least not at Harvard. And some kid named Aaron Greenspan on campus had gotten in trouble a few months before for getting kids to join an info-sharing bbs that had used their Harvard e-mails and IDs as passwords. Then the Greenspan kid had gone on to develop something called houseSYSTEM that had some social elements involved in it. Greenspan had even added a Universal House Facebook into his site, which Mark had checked out; hardly anyone had paid any attention to it, as far as Eduardo knew.

Friendster wasn't exclusive, the way Mark was describing his idea. And Greenspan's site wasn't particularly slick, and wasn't about pictures and profiles. Mark's idea was really different. It was about moving your real social network onto the Web.

"Isn't the school working on some sort of online facebook?"

Eduardo also remembered reading in the *Crimson* article on Facemash that the university actually did have plans in place to make some sort of universal online student picture site; other schools already had them, a sort of repository for school photos and such.

"Yeah, but what they're doing isn't interactive or anything. It's not what I'm talking about at all. And the Facebook is a pretty generic name. I don't think it matters where else it's being used."

Interactive—an interactive social network. It sounded pretty compelling. It also sounded like a lot of work—but Eduardo wasn't a computer expert. That was Mark's department. If Mark felt he could build such a site—well, then he could.

And it seemed like Mark had already done a lot of thinking about the idea—it was pretty developed, at least in his mind. Eduardo realized it was more than just Facemash—it also incorporated some of the stuff Mark had done with Course Match—where kids could see what classes other kids had taken. Friendster, of course, must have fed into it as well; certainly Mark had checked out the site, hadn't everybody?

Mark must have taken all those things, combined them in his head—and then taken it all a step further. Eduardo wondered when the genius moment had struck—while Mark was home, in Dobbs Ferry, over the holidays? While he was sitting alone in his dorm room, staring at his computer screen? In class?

The one place he was pretty sure Mark didn't have the stroke of genius was while hanging out with the Winklevoss twins. Mark had described the dinner meeting in full detail, as well as the site the Winklevosses thought Mark was working on for them. The way Mark had described it, it was little more than a dating Web site, a place for guys to try to get laid. A sort of highbrow Match.com.

As far as Eduardo knew, Mark hadn't actually done any work for the twins. He'd looked at their site, thought it through—and decided it wasn't worth his time. In fact, he'd scoffed at it, saying that even his most pathetic friends knew more about getting people interested in a Web site than Divya and the Winklevosses. Anyway, he was too busy with classes to spend time playing with a dating site just to impress a couple of Porc jocks. Though Eduardo was pretty sure Mark had continued to converse with them via e-mail and even phone calls, for God knows what reasons. Probably, because they were who they were—and Mark was who he was.

Eduardo was certain the Winklevoss twins had completely misread his friend. They'd probably looked at him and seen a geek who would jump at the chance to "rehabilitate" his image by building their Web site for them. But Mark didn't want to rehabilitate anything. Facemash had gotten him in trouble—but it had also shown the world exactly what Mark had wanted to show—that he was smarter than everyone else. He'd beaten Harvard's computers, then he'd beaten the ad board.

Certainly, Mark saw himself as leagues beyond the Winklevoss twins. Who were they to try to harness his abilities? Just a couple of jocks who thought they ruled the world. Maybe they did rule the social world, but in the land of Web sites and computers—Mark was king.

"I think it sounds great," Eduardo said. The restaurant had receded into the background, now, and all he could see was Mark's passion for this new project. Eduardo wanted to be involved. Obviously, Mark wanted his involvement as well. Otherwise, he would have gone to his roommates. One of them, Dustin Moskovitz, was a computer genius, maybe as good at coding as Mark. Why hadn't Mark gone to him first? There had to be a reason.

"It is great. But we're going to need a little start-up cash to rent the servers and get it online."

And there it was. Mark needed money to get his site going. Eduardo's family was wealthy—and more than that, Eduardo himself had money, the three hundred thousand dollars he'd made trading oil futures. The profits that had come from his obsession with meteorology, and the algorithms that had enabled him to predict hurricane patterns. Eduardo had money, Mark needed money—maybe it was as simple as that. But Eduardo wanted to believe there was even more to it.

What Mark was talking about was a social site. Mark had no social skills to speak of, and really no social life either. Eduardo had just become a member of the Phoenix. He was starting to branch out, meet girls. Sooner or later, he was probably even going to get laid. Of Mark's friends, who else could Mark have turned to? Eduardo was certainly the most social of the bunch.

"I'm in," Eduardo said, shaking Mark's hand across the table. He could provide money, and advice. He could help guide this project in a way that even Mark probably couldn't. Mark wasn't a business-minded kid. Hell, he'd turned down seven figures from Microsoft in high school!

Eduardo had grown up in a world of business. With this idea, perhaps he could show his father how much he had already learned. The head of the Harvard Investment Association was one thing; creating a popular Web site would be another entirely.

"How much do you think we'll need?" Eduardo asked.

"I think a thousand dollars to start. The thing is, I don't really have a thousand dollars at the moment, but if you put up what you can right now, we can get this thing off the ground."

Eduardo nodded. He knew that Mark wasn't rich; but Eduardo could have a thousand bucks ready in less than twenty minutes. All it would take was a short trip over to the nearest bank.

"We'll split the company seventy-thirty," Mark suddenly volunteered. "Seventy percent for me, thirty percent for you. You can be the company's CFO."

Eduardo nodded again. It sounded fair. It was Mark's idea, after all. Eduardo would finance it, and make the business decisions. Maybe they'd never make any money off the thing—but Eduardo had a feeling it was too good an idea to just fizzle away.

Kids all over campus were trying to build Web sites. Not just the Winklevosses and that Greenspan kid. Eduardo personally knew about a dozen other students who were trying to launch online businesses from their dorm rooms. Lots of them had social aspects like the Winklevosses' site—but none of the ones that Eduardo had heard of were anything as cool-sounding as Mark's idea. Simple, sexy, and exclusive.

The Facebook had all the elements of a successful Web site. A simple idea, a sexy function—and an exclusive feel. Like a Final Club, except online. It was the Phoenix, but you could join from the privacy of your own dorm room. And this time, Mark Zuckerberg wasn't going to just get punched. He was going to be made president.

"This is going to be really interesting." Eduardo grinned.

Mark grinned right back at him.

CHAPTER 12 | JANUARY 14, 2004

The door was huge and painted pitch-black; right across Mass Ave. from an even larger, more ominous stone gate—complete with iron bars, ornate masonry, and a great limestone boar's head carved into its arched pinnacle. There was no way any freshman who walked through that gate, glanced across the street toward that door, didn't feel at least a tingle of curiosity—if not outright paranoia. The building itself might have been nondescript, reddish bricks rising up four floors above an austere clothing store; but 1324 Mass Ave. was a place of Harvard myth and legend—an address intertwined with the secret history of the university itself.

At the moment, Tyler Winklevoss, his brother Cameron, and their best friend Divya, were seated on a green leather, L-shaped couch just inside that black door, in a small, rectangular parlor known only as the Bicycle Room. If it had just been Tyler and Cameron, they would have been sequestered on a higher floor; but the wooden, green-carpeted staircase that led up into the century-old building was off-limits to

Divya. Divya had never been invited up those winding, narrow stairs—and he never would be.

The Porcellian was a place of rules; for more than two centuries, the Porcellian had sat atop the Final Club hierarchy, the highest rung of a social order that had trained generations of the best and brightest the country had ever educated. It was, arguably, the most elite and secretive club in America—comparable to the Skull and Bones at Yale. Founded in 1791, named in 1794 in honor of a bacchanalian pig roast that the graduating members had thrown for themselves—feasting on a pig, the story goes, that one member had brought to classes with him, hiding the porcine pet in a window box whenever a professor came near—the Porcellian was the ultimate old boys' network on a campus that had defined the term.

The clubhouse—"the old barn," as the members referred to it—was a place of legend and history. Teddy Roosevelt had been a Porc, along with many members of the Roosevelt clan; FDR had been rejected from the club, and had called the incident "the greatest disappointment of his life." The Porcellian's motto—*dum vivimus, vivamus,* "while we live, let's live"—did not apply simply to a member's experience at college, but well after, as he went out and made his way into the world. Porcs were meant to become masters of the universe; there was even an urban myth on campus that if a Porc member hadn't made his first million by the age of thirty, the club simply gave it to him.

Whether or not that was true, Tyler, Cameron, and Divya hadn't come to the Bicycle Room to contemplate the path to their first million; they were there to commiserate, because suddenly success seemed more distant than ever.

The reason for their frustrated state had a name: Mark Zuckerberg.

For two months, since that seemingly wonderful meeting of the minds in the Kirkland House dining hall, the kid had been telling them that their partnership in the Harvard Connection was going great. He'd looked over their computer code, studied what they'd already built of the site, and was ready to do his part to get it up and running.

Fifty-two e-mails between Mark, the Winklevosses, and Divya, a half-dozen phone calls—and always, the kid had seemed as thrilled and excited about the project as he had been during that first dinner meeting. His e-mails had been like a work log to the Winklevosses, progress reports that they thought indicated the programming was moving steadily along, if a little slower than expected:

> **Most of the coding done, It seems like everything is working.**
> **Got some class work I have to get done, be back at it soon.**
> **I forgot to bring my charger home with me for Thanksgiving.**

But by the end of the seventh week, when no real progress had been forthcoming—no code e-mailed to them or added to the site—Tyler had begun to get a little anxious. Things were taking way too long. He'd thought they'd have been ready to get the site launched by the end of the holidays. So he'd had Cameron send the kid an e-mail, asking if he could finish the job soon. Mark had responded almost immediately, but the response had been a request for more time:

> **Sorry it's taken me a while to get back to you. I'm completely swamped with work this week. I have three programming projects and a final paper due Monday, as well as a couple problem sets due Friday.**

But in the same e-mail, Mark had let them know that he was still working on the site as much as he could:

As far as the site goes for now, I've made some of the changes, although not all of them, and they seem to be working on my computer. I have not uploaded them to the live site yet though.

And then he'd added something that had caused Tyler a little concern, because it seemed out of the blue, considering how upbeat Mark had seemed until then:

I'm still a little skeptical that we have enough functionality in the site to really draw the attention and gain the critical mass necessary to get a site like this to run. And in its current state, if the site does get the type of traffic we're looking for, I don't know if we have enough bandwidth from the ISP you're using to handle the load without some serious optimization, which will take a few more days to implement.

It was the first time Mark had mentioned anything about the site not having "functionality"; up until then, he had seemed thrilled with their ideas, and had agreed that it would be a great success.

After that e-mail, Tyler had been insistent, putting the pressure on the kid to meet with them. He had hoped that the site would be ready to go online by now, and every day they wasted was a day that someone else could beat them to the punch—get a good similar site up and running. Tyler and Cameron were seniors, they wanted to see their project happen as soon as possible. But Mark had kept postponing, claiming he had too much schoolwork to schedule anything.

It wasn't until that very night, just a few hours before the Winklevosses and Divya had crossed through the Porcellian gate—

donated to Harvard by the club in 1901—and entered that pitch-black door, that Mark had finally acquiesced to a brief get-together in the Kirkland dining hall.

At first, when Tyler, Cameron, and Divya had joined the kid at the same back table, it had seemed just like before; the kid complimented them on their ideas, told them how great he thought the Harvard Connection was going to be—but then, out of nowhere, he'd started to hedge a bit, explaining that he didn't have time to get much done right away, that he had a lot of other projects that were taking up a lot of his free hours. Tyler assumed he was talking about projects for his computer classes—but Mark was being very vague, very unclear.

He also had brought up a few problems he was having with the Harvard Connection that he'd never mentioned before; that there was some "front-end stuff" that needed to be done, and that he wasn't good at that. By "front-end stuff," Tyler assumed he meant the visual aspects of the front page, which seemed strange, because that was exactly what Mark had shown himself to be very talented at with the Facemash debacle.

Then Mark had gotten even more confusing—stating that some of the work he still needed to do to get the site live was "boring," stuff he wasn't interested in doing. He again reiterated that the site was lacking "functionality." That they were going to need more server capacity.

Tyler had suddenly gotten the feeling that the kid was trying to deflate their balloon; where he had been enthusiastic before, now he was trying to tell them that it just wasn't that exciting to him.

Tyler had wondered—maybe the kid was just burning out a bit. He was working hard, with all his classes, and Tyler knew from Victor that engineers had a tendency to get like that, a little burned out, a little tired, a little testy. The kid's excuses seemed pretty hollow, that was for sure. Server problems? So they'd get more servers. Front-end issues? Anyone could design the front end. Maybe he just needed some time left alone—then he'd get right back to work. Maybe by February he would be enthusiastic again.

Still, it was extremely frustrating, and Tyler, Cameron, and Divya had come out of the meeting utterly depressed. After all those weeks of telling them that everything was going along fine, now Mark was telling them it wasn't ready to go, that there were some real issues he was dealing with, that he wasn't that excited anymore. No real explanation other than schoolwork, nothing more than a lame apology—and another two months wasted.

It was beyond disappointing. Tyler had really thought the site would be ready to be launched by now. He'd really thought the geeky kid had gotten their project, understood the possibilities. The kid had seen what they'd already done, had agreed that it would be easy to finish—maybe ten, fifteen hours of work for a competent computer programmer—but now all this garbage about front ends and server capacity.

It didn't make any sense. Tyler had ultimately decided that the best course of action was to leave the kid alone for a few weeks. Maybe he'd be back to his old self.

"And if he doesn't get it together in a few weeks?" Divya asked as they sat on the couch in the Bicycle Room. They could hear cars driving by on Mass Ave. on the other side of the black door. If Tyler and Cameron had gone upstairs, they could have watched the traffic through a mirror designed specifically so that nobody could see them watching; but Tyler had never been much of a voyeur. He wanted to participate, to be a part of things, to move forward. He hated being stalled, just watching as the rest of the world went by.

Tyler shrugged. He didn't want to get ahead of himself—but maybe they had read the kid wrong. Maybe Mark Zuckerberg wasn't the entrepreneur Tyler had thought he was. Maybe Zuckerberg was just another computer geek without any real vision.

"If that happens," Tyler glumly responded, "we have to find ourselves a new programmer. One that understands the big picture."

Maybe Mark Zuckerberg didn't get it at all.

CHAPTER 13 | FEBRUARY 4, 2004

Eduardo had been standing in the empty hallway in Kirkland House a good twenty minutes before Mark finally burst out of the stairwell that led down toward the dining hall; Mark was moving fast, his flip-flops a blur beneath his feet, the hood of his yellow fleece hoody flapping behind his head like a halo in a hurricane. Eduardo, watching his friend careening forward, crossed his arms against his chest.

"I thought we were supposed to meet at nine," Eduardo started, but Mark waved him off.

"Can't talk," he mumbled as he dug his key out of his shorts and went to work on the doorknob.

Eduardo took in his friend's wild hair and even wilder eyes.

"You haven't slept yet, have you?"

Mark didn't respond. The truth was, Eduardo was pretty sure Mark hadn't slept much in the past week. He had been working round the clock, light to dark to light. He looked beyond exhausted, but it didn't matter. At the moment, nothing mattered to Mark. He was in that pure laser mode that every engineer understood. He refused to suffer any distraction, anything that could jar the single thought loose.

"Why can't you talk?" Eduardo continued, but Mark ignored him. Finally, the keys clicked and Mark got the door open, diving inside. His flip-flops caught in a pair of jeans that were balled up on the floor, and he momentarily lost his balance, spinning past a cluttered bookshelf and a small color television. Then he was back on his feet, and still moving forward. He launched himself into his bedroom, beelining straight to his desk.

The desktop computer was on, the program open—and Mark went straight to work. He didn't seem to hear Eduardo plodding across the room behind him. He hit the keys furiously, his fingers moving like they were possessed.

He was adding a final touch, Eduardo assumed, because all the debugging had been finished by three, and most of the design and coding were already finished. The only thing that had been missing had been a function that Mark had been mulling over for nearly a day.

He'd been playing around with the features of the site, trying to keep the design as simple and clean as possible, while providing enough pizzazz to draw a viewer's attention. It wasn't just voyeurism that was going to drive people to use thefacebook. It was the interactivity of that voyeurism. Or, to put it more simply, it was going to mimic what went on at college every day—the thing that drove the college social experience, drove people to go out to the clubs and bars and even the classrooms and dining halls. To meet people, socialize, converse, sure—but the catalyst of it all, the burning engine behind those social networks, was as simple and basic as humanity itself.

"That looks pretty good," Eduardo said, reading over Mark's shoulder. Mark nodded, mostly to himself.

"Yes."

"No, I mean, that's great. That looks great. I think people are going to really respond to this."

Mark rubbed a hand through his hair, leaning back in his chair. The page was open to the inside of the site—a mock-profile page, what people would see after they registered and added their personal info. There was a picture near the top—whatever picture you wanted to add. Then a list of attributes on the right side—year you were in at college, your major, your high school, where you came from, clubs you were a member of, a favorite quote. Then a list of friends—people you could add yourself, or invite to join. A "poke" application, that allowed you to poke other people's profiles, letting them know that you were checking them out. And, in big letters, your "Sex." What you were "Looking For." Your "Relationship Status." And what you were "Interested In."

That was the genius of it, that the addition was going to make this all work. *Looking For. Relationship Status. Interested In.* Those were the résumé items that were at the heart of the college experience. Those three concepts, in a nutshell, defined college life—from the parties to the classrooms to the dorms, that was the engine that drove every kid on campus.

Online, it would be the same; the thing that would drive this social network was the same thing that drove life at college—*sex.* Even at Harvard, the most exclusive school in the world, it was all really about sex. Getting it, or not getting it. That's why people joined Final Clubs. That's why they chose certain classes over other ones, sat in certain seats at the dining hall. It was all about sex. And deep down, at its heart, that's what thefacebook would be about, in the beginning. An undercurrent of sex.

Mark hit more keys, changing the page to the opening screen that you'd see when you went to thefacebook.com. Eduardo took in the dark blue band across the top, the slightly lighter blue "register" and "login" buttons. It was extremely simple- and clean-looking. No blinking lights, no annoying bells. It was going to be all about the experience—nothing flashy, nothing overwhelming or frightening. Simple and clean:

[Welcome to Thefacebook]

Thefacebook is an online directory that connects people through social networks at colleges.

We have opened up Thefacebook for popular consumption at Harvard University.

You can use Thefacebook to:

- **Search for people at your school**
- **Find out who are in your classes**
- **Look up your friends' friends**
- **See a visualization of your social network**

To get started, click below to register. If you have already registered, you can log in.

"So to log on," Eduardo said, his hovering shadow covering most of the screen. "You need a Harvard.edu e-mail, then you choose a password."

"Correct."

The Harvard.edu e-mail was key, in Eduardo's mind; you had to be a Harvard student to join the site. Mark and Eduardo knew that exclusivity would make the site more popular; it would also enhance the idea that your info would remain in a closed system, private. Privacy was important; people wanted to have control of what they put onto the Web. Likewise, choosing your own password was integral; that Aaron Greenspan kid had gotten into so much trouble for having students use their Harvard ID numbers and system passwords to log onto his site. Mark had even e-mailed with him about his experience, the trouble he'd had with the ad board. Greenspan had immediately tried to get Mark to partner up with him—just like the Winklevoss twins and their Harvard Connection dating site. Everyone wanted a piece of Mark, but Mark didn't need anyone else. Everything he needed was right in front of him.

"And what's that, at the bottom?"

Eduardo was leaning forward, squinting to read a small line of print.

A Mark Zuckerberg production.

The line would appear on every page, right there at the bottom of the screen. Mark's signature, for everyone to see.

If Eduardo had a problem with that, he didn't say anything. And why should he? Mark had been working so hard—the hours must have blended together in one blur of pure programming. He had barely eaten, barely slept. It seemed like he had missed almost half of his classes, and had probably been in real danger of screwing up his grade-point average. In one class—one of his stupid Cores called Art in the Time of Augustus—he'd supposedly fallen so far behind that he'd almost forgotten about an exam that was going to be worth a large percentage of his overall grade. He'd had no time to study for the damn thing—so he'd reportedly figured out a unique way of dealing with the situation. He'd created a quick little Web site where he posted all the artwork that was going to be on the exam and invited people in the class to comment—effectively creating an online crib sheet for the test. He'd essentially gotten the rest of the class to do the work for him—and he'd aced the exam, saving his grade.

And now, sitting here in front of Mark's creation, it seemed like all the work had paid off. The Web site was pretty much done. They had registered the domain name—thefacebook.com—a couple of weeks ago, January 12. They'd booked the servers—around eighty-five bucks a month—from a company in upstate New York. They'd take care of any Web traffic and maintenance; Mark had obviously learned his lesson from the Facemash incident, he didn't need any more frozen laptops. The servers could handle a pretty large amount of traffic, so there wouldn't be any problems with the site freezing up, even if the thing was as popular as Facemash had been. Everything was in place.

Thefacebook.com was ready to roll.

"Let's do this."

Mark pointed to his laptop, which was open on the desk next to his desktop computer. Eduardo moved beside him, hunching over the laptop keyboard, his sloped shoulders curved inward as he attacked the keys. He quickly opened his e-mail address book and pointed to a bunch of names grouped together near the top.

"These guys are all members of the Phoenix. If we send it to them, it will get spread around pretty fast."

Mark nodded. It had been Eduardo's idea to go to the Phoenix members first. They were the social stars on campus, after all. And thefacebook was a social network. If these kids liked it, and sent it on to their friends, it would spread pretty fast. And these Phoenix guys knew lots of girls. If Mark had simply tried to send it out to his own e-mail list, it would bounce around the computer science department. And the Jewish fraternity, of course. Certainly it wouldn't get to many—if any—girls. And that would be a problem.

The Phoenix was a much better idea. That—along with the Kirkland House e-mail list, which Mark had legal access to, as a member of the house—would get this thing started right.

"Okay," Eduardo said, with a quaver in his voice. "Here we go."

He wrote a simple e-mail, just a couple lines of text, introducing the site, and linked in thefacebook.com. Then he took a deep breath, and hit the key—sending out the mass e-mail with a single stroke of his finger.

It was done. Eduardo closed his eyes, imagining the tiny packets of information ricocheting out into the world, whizzing down copper tubing and bouncing off of orbiting satellites, ripping through the ether, tiny bursts of electric genius leaping from computer to computer like synaptic flashes in a vast, worldwide nervous system. The Web site was out there.

Live.

Alive.

Eduardo put a hand on Mark's shoulder, startling him.

"Let's get a drink! It's time to celebrate!"

"No, I'm going to stay here."

"You sure? I hear there are some girls coming over to the Phoenix later. They sent the Fuck Truck for 'em."

Mark didn't respond. At the moment, Eduardo could tell from Mark's expression that he was a distraction, like the sound of the radiators near the wall or the traffic in the street down below his little window.

"You're going to just stay here and stare at the computer screen?"

Again, Mark didn't answer. He was bobbing a little behind the computer, davening, even.

It was a strange sight, but Eduardo obviously decided not to judge his awkward friend. And why should he? Mark had been working round the clock to get thefacebook ready for this launch. If he wanted to sit by himself and stare, he'd earned the right.

Eduardo backed away from him, crossing the small bedroom in

near silence. Then he paused at the doorway, tapping the door frame with his outstretched fingers. Mark still didn't turn around. Eduardo shrugged, turned, and left the kid alone with his computer.

Mark sat there enveloped in silence, lost in his own reflection as it danced across the screen.

CHAPTER 14 | FEBRUARY 9, 2004

Tyler was in the zone. Eyes closed, muscles rippling across his back, chest heaving, quads and triceps and forearms burning, fingers white against the oars. The blades sliced in and out of the water without so much as a ripple, mimicked exactly by Cameron's pair just a few feet behind—utterly in sync, again and again and again. Tyler could almost hear the cheering of the fans who packed the banks of the Charles, he could almost see that bridge coming closer and closer and closer—

"Tyler! You've got to see this!"

And it all came crashing down. His oars wobbled in his grip and the water splashed upward, soaking his sweatshirt and shorts. His eyes whipped open—and he didn't see the banks of the Charles flashing by. He saw the interior of the Newell Boathouse, home to Harvard's crew team since 1900. He saw a cavernous, hall-like room, walls lined with ancient crew memorabilia—oars and hulls and sweatshirts, framed black-and-white photos and shelves full of trophies. And he saw the angry-looking Indian kid standing a few feet in front of him, holding up a copy of the *Harvard Crimson*.

Tyler blinked, then let his oars down and wiped water from his cheeks. He glanced back at his brother, who had also stopped rowing. The two of them were sitting in one of Newell's two state-of-the-art "tanks"—the indoor rowing pools consisting of a concrete-walled eight-man "hull" surrounded on both sides by huge ditches of rowable water. Tyler knew that they probably looked ridiculous, sitting there in the tank, soaking wet—but Divya wasn't smiling, that was for sure. Tyler looked at the *Crimson* in his friend's hands, and rolled his eyes.

"What is it with you and that newspaper?"

Divya held it out toward him, so angry that his hands were shaking. Tyler shook his head.

"You read it. I'm soaking wet. I don't want to get newsprint all over me."

Divya exhaled, exasperated, then opened the paper and started reading:

" 'When Mark E. Zuckerberg '06 grew impatient with the creation of an official universal Harvard facebook, he decided to take matters into his own hands'—"

"Hold on," interrupted Cameron. "What the hell is that?"

"Today's paper," Divya responded. "Listen to this: 'After about a week of coding, Zuckerberg launched thefacebook.com last Wednesday afternoon. The website combines elements of a standard House face book with extensive profile features that allow students to search for others in their courses, social organizations and Houses.' "

Tyler coughed. Last Wednesday afternoon? That was four days ago. He hadn't heard anything about this Web site—but then again, he and his brother had been going at their training like animals. He'd barely checked his e-mail in that time.

"This is crazy," he said. "He launched a Web site?"

"Oh yeah," Divya said. "Here, they quote him right in the article. ' "Everyone's been talking a lot about a universal face book within Harvard," Zuckerberg said. "I think it's kind of silly that it would take the University a couple of years to get around to it. I can do it better than they can, and I can do it in a week." ' "

He can do it in a week? In Tyler's view, he had been putting Tyler and the Harvard Connection off for two months, saying that he didn't have the time to program the site, that he had too much going on with his classes and the holidays. Christ, Tyler thought, Mark had lied straight to their faces! In fact, Cameron had sent him an e-mail barely two weeks before, asking Mark's advice on some design issues for the Harvard Connection—and he'd never responded. They had assumed he was still too bogged down in schoolwork.

Tyler thought, he'd had time to make his own fucking Web site—but he hadn't had time to give them ten hours of coding?

"It gets worse. 'As of yesterday afternoon, Zuckerberg said over 650 students had registered to use thefacebook.com. He said that he anticipated that 900 students would have joined the site by this morning.' "

Holy shit. That couldn't be true. Nine hundred students had signed up to his Web site in four days? How was that possible? Zuckerberg didn't know nine hundred people. He didn't know four people, as far as Tyler could tell. In Tyler's view, the kid had no friends. He had no social life. How the hell had he launched a social Web site and gotten that kind of response in four days?

"I checked the site out as soon as I read this. It's true, the thing is really exploding. You have to have a Harvard e-mail, and then you get to upload your picture, and personal and academic info. You can search for people according to interests, and then when you find your friends, you make a network out of them."

Tyler felt his hands tightening. It wasn't the same as the Harvard Connection—but in his mind, it wasn't that different, either. The Harvard Connection was going to be about searching out people based on interests. And it was going to be centered on the domain of Harvard. Had Zuckerberg just taken their idea and run with it? Could it be a coincidence—had he been meaning to work on their site, but had just gotten carried away with his own?

No, it didn't seem right. To Tyler, it seemed like . . . theft.

"From what I hear, he got some financing from one of his buddies, a Brazilian kid named Eduardo Saverin. He's in the Phoenix, made some money trading stocks over the summer. Now he's part owner of the site."

"Because he paid for it?"

"I guess."

"Why didn't Mark come to us?"

Mark assuredly knew that the Winklevosses had money; he must have known they were in the Porc, and everyone knew what that meant. If he'd needed cash to start a site, he could easily have mentioned it to Tyler or Cameron. *Unless the thing he needed cash for was something he had stolen from them.* Unless the Web site he was working on had to be kept secret from them, because it was too similar to what they had hired him to do. Well, not hired, exactly—they had never talked about paying him, just that he'd benefit if they benefited.

There had been no contract, no paperwork, nothing but a handshake here and there. *Fuck.* Tyler lowered his head, staring at the blue-green water in the rowing tank. Why hadn't they written something up, even just some bullshit one-pager—you do this, we'll do that—something simple. Instead, they'd just trusted the kid. Now it seemed to Tyler like he'd fucked them over. He'd stalled them, led them on, then launched his own site with similar features.

"Here's the best part," Divya said, back to reading from the *Crimson.* " 'Zuckerberg said that he hoped the privacy options would help to restore his reputation following student outrage over facemash .com, a website he created in the fall semester.' "

Tyler slammed one of the oars with his palm, sending a plume of water splashing up out of the tank. Almost the exact words of his pitch to Mark—that the Harvard Connection would restore his reputation—and Mark had used them, right there in the *Crimson.* It was almost as if Mark was mocking them.

In Tyler's view, he'd strung them along for two months, right through the holidays and the winter reading period—all the while, working on his own Web site. Then he'd blown them off, and, barely two weeks later, launched his own site—thefacebook.com, stealing their thunder, and in Tyler's mind, the essence of their idea.

"What are we going to do?" Cameron asked.

Tyler wasn't sure. But he knew he couldn't just let it happen. He couldn't let that fucking weasel get away with it.

"First, we're going to make a phone call."

▶ ▶ ▶

Tyler's mind worked furiously as he held the phone hard against his ear. He was standing in his dorm room in Pforzheimer, still soaking wet from a hasty shower, a towel over his shoulders and a pair of sweatpants loose around his waist. Cameron and Divya were at his desk a few feet away, surfing through Zuckerberg's site on Tyler's desktop computer. Every time Tyler glanced toward them, and saw that blue-bordered screen, his cheeks heated up, and fire sparked behind his eyes. This wasn't right, damn it. This wasn't fair.

His dad finally answered on the third ring. There was no one in the world Tyler respected more. His father, a self-made multimillionaire, ran one of the most successful consulting companies on Wall Street. If anyone was going to know how to handle a difficult situation like this, it was him.

Tyler spoke quickly into the phone, explaining exactly what had happened. His dad knew all about the Harvard Connection; after all, they'd been working on the site since December of 2002. Tyler gave him the background of their relationship with Zuckerberg, then told him what he'd read in the *Crimson*—and what he, Cameron, and Divya had seen for themselves, logging into thefacebook.com.

"There are things that seem real similar, Dad."

The key, to Tyler, was the setting, the exclusivity of it, that really separated what Mark had made from social network sites like Friendster. You had to have a Harvard e-mail to enter Mark's site—and that had been their idea, too, to launch a Harvard-centric social Web site. The very idea of making everyone who joined have an .edu e-mail address was completely innovative, and potentially very important to the initial success of the site. It was sort of a screening process that kept the thing exclusive and safe. Maybe a lot of the features Mark had put in thefacebook.com were different—but the overall concept, to Tyler, seemed too similar.

Mark had met with them three times. They had exchanged fifty-two e-mails—all of which were still on Cameron, Tyler, and Divya's computers. Mark had looked at their code—which they could prove. He'd seen what Victor had already done, and had talked to them at length about what they planned to do.

"It isn't about money," Tyler concluded. "Who knows if either of our sites are ever going to make any money. But this just isn't right. It isn't fair."

This wasn't how the world was supposed to work. Tyler and Cameron had grown up believing that order mattered. Rules mattered. You worked hard, you got what you deserved. Maybe in Mark's hacker world—his computer-geek worldview—things were different. You just did whatever the hell you wanted, you launched prank sites like Facemash, you hacked into Harvard's computers, you thumbed your nose at authority and mocked people right in the pages of the *Crimson*—but that simply wasn't acceptable.

That wasn't Harvard. Harvard was a place of order. Wasn't it?

"I'm going to put you on with my in-house counsel," Tyler's dad said.

Tyler nodded, slowing his breathing, forcing calm into his veins. A lawyer, that's exactly what they needed. They needed to go over their options with a professional, see what could be done.

Maybe it wasn't too late. Maybe, just maybe, they could still make this right.

CHAPTER 15 | AMERICAN IDOL

From up above, the man looked tiny and hunched behind the podium, his face just a little too close to the microphone, and his lanky shoulders poked out at the corners of his formless beige sweater. His bowl haircut dribbled almost to his eyes, and his oversize glasses covered most of his splotchy face, obscuring any sense of expression or emotion; his voice reverberating through the speakers seemed a little too high and nasal, and sometimes it veered into a monotone drone, a single laryngeal note played over and over again until the words bled right into one another.

He was not a fantastic speaker. And yet, just his presence, the mere fact that he was standing there in the front of Lowell Lecture Hall with his pale hands flapping against the podium, his turkey neck bobbing up and down as he tossed pearls of monotone wisdom at the crowded room—it was beyond inspiring. The audience—made up mostly of engineering and computer geeks from the CS department and a few econ majors with entrepreneurial aspirations—hung on every nasal word. To the gathered acolytes, this was heaven, and the strange, bowl-cutted man at the podium was god.

Eduardo sat next to Mark in the back row of the balcony, watching as Bill Gates mesmerized the gathered crowd. Despite Gates's strange, almost autistic mannerisms, he managed to toss off a few jokes—one about why he'd dropped out of school ("I had a terrible habit of not going to classes") and certainly some pearls of wisdom—that AI was the future, that the next Bill Gates was out there, possibly in that very room. But Eduardo specifically saw Mark perk up when Gates answered a question from one of the audience members about his decision to leave school and start his own company. After hemming and hawing a bit, Gates told the audience that the great thing about Harvard was that you could always come back and finish. The way Mark seemed to smile when Gates said it made Eduardo a little nervous—especially considering how hard Mark had been working on simply keeping up with the demand of their nascent Web site. Eduardo would never drop out of school—it simply wasn't a possibility to him. In the first place, his father would throw a fit; to the Saverins, nothing was more important than education, and Harvard meant nothing if you didn't come out of there with a degree. Second, Eduardo understood that entrepreneurship meant taking risks—but only to a certain degree. You didn't risk your entire future on something until you figured out how it was going to make you rich.

Eduardo was so busy watching Mark watch Gates, he almost didn't hear the giggles coming from the seats right behind him; he might not have turned to look if the whispered voices that followed the laughter hadn't been decidedly female.

As Gates droned on, answering more questions from the packed crowd, Eduardo glanced over his shoulder—the seat behind him was empty, but from the row right behind the empty seat, he saw two girls smiling and pointing. The girls were both Asian, pretty, and a little overly made up for a lecture like this. The taller of the two had long sable hair pulled back in a high ponytail and was wearing a short skirt and a white shirt open one button too far down the front; Eduardo could see wisps of her red lace bra, wonderfully offset by her tan, smooth skin. The other girl was in an equally short skirt, with a black leggings combo that showed off some impressively sculpted calves. Both had bright red lipstick and too much eye shadow, but they were damn cute—and they were smiling and pointing right at him.

Well, at him and Mark. The taller of the girl leaned forward over the empty seat and whispered in his ear.

"Your friend—isn't that Mark Zuckerberg?"

Eduardo raised his eyebrows.

"You know Mark?" There was a first time for everything.

"No, but didn't he make Facebook?"

Eduardo felt a tingle of excitement move through him, as he felt the warmth of her breath against his ear, as he breathed in her perfume.

"Yeah. I mean, Facebook, it's both of ours—mine and his."

People had dropped the *the,* were pretty much calling it Facebook all over campus. And even though it had only been a couple of weeks since they'd launched the site, it already felt like everyone was on it—because, well, everyone at Harvard *was* on it. According to Mark, they had now signed up five thousand members. Which meant that almost 85 percent of the university undergraduates had put up a Facebook profile.

"Wow, that's really cool," the girl said. "My name is Kelly. This is Alice."

Other people in the girls' row were looking now. But they didn't seem angry that the whispers were interrupting their enjoyment of

Bill Gates. Eduardo saw someone point, then another kid whisper something to a friend. Then more pointing—but not at him, at Mark.

Everyone knew Mark now. The *Crimson* had made sure of that—printing article after article about the Web site, three already in the past week. Quoting Mark about the Web site, even printing his picture. Nobody had interviewed Eduardo—and the truth was, he was happy about that. Mark wanted the attention; Eduardo just wanted the benefits that came with attention, not the attention itself. This was a business they'd created, and getting it out there was important, but Eduardo didn't want to be a celebrity because of it.

And it was beginning to look like becoming a celebrity was a real possibility. Though thefacebook had been up and running only for a short time, it was really changing life at Harvard. It was insinuating itself into everyone's routine: you got up, you checked your Facebook account to see who had invited you to be their friend—and which of your invites had been accepted or rejected. Then you went about your business. When you got home, if there was a girl you saw in one of your classes—or even just somebody you'd met in the dining hall—you simply searched for her on Facebook, then invited her to be your friend. Maybe you added some little message about how you'd met, or what you saw in her listed interests that jibed with one of your own. Or maybe you just invited her cold, no message, just to see if she knew you existed. When she opened her account, she'd see your invite, look over your photo, and maybe accept your invitation.

It was really such an amazing tool, lubricating the social scene—making everything happen so much faster. But it wasn't a dating Web site—the way Eduardo saw Friendster. For all its hype as a social network, Friendster—and MySpace, which was just beginning to catch fire nationwide—was really just about searching through people you didn't know and trying to hook up with them. The difference was, on Facebook you already *knew* the people you invited to be your friends. You might not know them well, but you knew them. They were classmates—or friends of friends, members of a "network" that you could join, or be asked to join, by people you knew who were already members.

That was the genius of it all. Mark's genius, really, but Eduardo felt he was a part of that as well. He'd put up the money for the servers—but he'd also had a hand in discussing some of the attributes of the site, the ideas behind some of the simplified structure.

What neither he nor Mark had known when they started the damn thing was how addictive Facebook was. You didn't just visit the site once. You visited it every day. You came back again and again, adding to your site, your profile, changing your pictures, your interests, and most of all, updating your friends. It really had moved a large portion of college life onto the Internet. And it really had changed Harvard's social scene.

But that didn't make it a business yet—just a highly successful novelty. Eduardo had some ideas about that, and after the lecture, he and Mark were going back to Mark's room to discuss them. The main thing he wanted to push Mark into understanding was that it was time to start chasing advertising dollars. That's how they would monetize Facebook, through ads. Eduardo knew it was going to be a tough sell; Mark wanted to just keep it as a fun site, not try to make any money off of it yet. But then again, he was the kid who had turned down a million bucks in high school. Who knew if he'd ever want to monetize Facebook?

Eduardo had a different worldview. Facebook was costing them money. Not much, just the cost of the servers, but as more people joined in, surely those costs would go up. The thousand dollars Eduardo had put into the Web site wasn't going to last forever.

Until the company had some sort of profit model, until they could figure out how to make money off of it—it was still just a novelty. Its value was certainly going up—but to turn that value into cash, they needed advertisers. They needed a business model. They needed to sit down and hash it all out. Most of all, Mark needed to let Eduardo do what he did best—think big.

"Very nice to meet you," Eduardo finally whispered back to the girls, who giggled again. The taller one—Kelly—leaned even closer, her lips almost touching his skin.

"Facebook me when you get home. Maybe we can all go out for a drink later."

Eduardo felt his cheeks blush. He turned back to Mark, who was looking at him now. Mark had obviously noticed the girls, but he didn't even try to talk to them. He raised his eyebrows for a second—then turned back toward Gates, his idol, and forgot all about them.

▶ ▶ ▶

It wasn't until two hours later, when Eduardo and Mark were finally ensconced in the overheated warmth of Mark's Kirkland dorm room—Eduardo absentmindedly picking through a stack of computer books that towered over the small color TV in the corner, while Mark lowered himself onto the ratty old couch in the center of the cheaply appointed common area, bare feet stretched out on the low coffee table in front of him—that Mark finally brought up the girls.

"Those Asian chicks were pretty cute."

Eduardo nodded, turning one of the books over, trying to make sense of the cover, which was covered in equations he knew he'd never understand.

"Yeah, and they want to meet us later tonight."

"That could be interesting."

"Could be—Mark, what the hell is this?"

A piece of paper had slipped out from beneath the computer book and had landed, facing up, on Eduardo's laced-up Italian leather shoes. Even from a stooped position, Eduardo could clearly make out the legal looking header and script; it was a letter, from some Connecticut law firm, and it looked really serious. It was addressed to Mark Zuckerberg, but from the first sentence alone Eduardo could see that it involved him, too. The words *TheFacebook* were hard to miss—as were the words *damages* and *misappropriated:*

From: Cameron Winklevoss
Sent: Tuesday, February 10, 2004 9:00 PM
To: Mark Elliot Zuckerberg
Subject: Important Notice

Mark,

It has come to our attention (Tyler, Divya and myself) that you have launched a website named TheFacebook.com. Prior to this launch, we had entered into an agreement with you under which you would help us develop our proprietary website (HarvardConnection) and do so in a timely manner (specifically noting that the window for launching our site was quickly closing).

Over the last three months, in breach of our agreement, and to our material detriment in reliance on your misrepresentations, fraud and/or other actionable behavior, for which we assert that damages are payable, you stalled the development of our website, while you were developing your own website in unfair competition with ours, and without our knowledge or agreement. You have also misappropriated our work product, including our ideas, thoughts, concepts and research.

At this time we have notified our counsel and are prepared to take action, based upon the above legal considerations.

We are also prepared to petition the Harvard University Administrative Board regarding your breach of ethical standards of conduct, as stated in the Student

Handbook. Please note that our petition will be based on your violation of the College's expectations of honesty and forthrightness in your dealings with fellow students, your violation of the standard of high respect for the property and rights of others, and your lack of respect for the dignity of others. Misappropriation is also actionable under these ethical rules, as well as at law.

We demand the following in order to put a temporary stay to these actions, until we have fully evaluated your website and what actions we may take:

1. Cease and desist all further expansion and updates of TheFacebook.com;
2. State in writing to us that you have done so; and
3. State in writing that you will not disclose to any third person our work product, our agreement, or this demand.
4. These demands must be met no later than 5pm Wednesday, February 11th, 2004

Notwithstanding your compliance with the above, we reserve the right to consider other action to further protect our rights and to recover damages against you. Your cooperation will prevent further violation of our rights and further damages.

Any failure to meet these demands will lead us to consider immediate action on both legal and ethical fronts. If you have any questions you are welcome to email me back or set up a meeting.

Cameron Winklevoss

Hardcopy also sent via University Mail

"I think they call it a cease-and-desist letter," Mark mumbled, leaning back against the couch, hands behind his head. "What were the girls' names? I liked the short one."

"When did you get it?" Eduardo said, ignoring Mark's question. He felt blood rushing into his head. He reached down, picked up the letter, read it over quickly. It looked pretty intense. It was full of accusations—and at the bottom, in clear words, it spelled out who was making the accusations. Tyler and Cameron Winklevoss, on behalf of their Web site the Harvard Connection. They were accusing Mark of stealing their idea, their code—and demanding that he and Eduardo shut down thefacebook or face legal action.

"A week ago. Right after we launched the site. They also sent an e-mail, a letter saying they were going to appeal to the school, too. That I had violated Harvard's code of ethics."

Jesus Christ. Eduardo stared at Mark, but, as usual, couldn't read anything from his blank expression. The Winklevosses were accusing Mark of stealing their idea? Their dating Web site? They wanted to shut thefacebook down?

Could they even do that? Sure, Mark had met with them, had e-mailed with them, had led them on. But he hadn't signed any contracts, and hadn't written any code. And to Eduardo, thefacebook seemed so different. Well, it was also a social Web site—but there were dozens—if not hundreds—of social Web sites. Hell, every computer science major on campus had a social Web site under development. That Aaron Greenspan kid had even called part of his networking portal "the facebook," or something like that. Did that mean they could all sue one another? Just for having similar ideas?

"I talked to a three-one at the law school," Mark said. "I sent a letter back. And another one to the school. Under that next book."

Eduardo reached for another computer book in the stack on the TV and found the second letter, this one written by Mark to the university. Eduardo skimmed it quickly, and was immediately surprised—and pleased—to see some real emotion in Mark's response to the Winklevosses' claims. Mark had told the university, in no uncertain terms, that thefacebook was not related to the tiny bit of work he'd done for the Winklevosses.

Originally, I was intrigued by the project and was asked to finish the Connect side of the website . . . After this meeting, and not before, I began working on TheFacebook, using none of the same code nor functionality that is present in Harvard Connection. This was a separate venture, and did not draw on any of the ideas discussed in our meetings."

Furthermore, Mark felt he had been fooled by the initial meeting, that the twins had misrepresented what they wanted him to do:

From the initiation of this project, I perceived it as a non-business oriented venture, with its primary purpose of developing an interested product to aid the Harvard community. I realized over time that my concept of the web site was not as it had initially been portrayed.

And what's more, Mark hadn't really led them on at all:

When we met in January, I expressed my doubts about the site (where it stood with graphics, how much programming was left that I had not anticipated, the lack of hardware we had to deal with, the lack of promotion that would go on to successfully launch the web site, etc.). I told you that I had other projects I was working on, and that those were higher priorities than finishing [your site]."

Mark had concluded that he was appalled to find himself "threatened" by the twins because of a few meetings in the Kirkland dining hall and some e-mail conversations he'd had with Cameron, Tyler, and Divya. And that he saw their claims as an "annoyance," something he was "shrugging off," that it was the kind of unabashed moneygrubbing you had to expect when you made something successful.

Which, of course, seemed a little over-the-top to Eduardo, considering that thefacebook wasn't making anyone any money—and the Winklevosses were hardly hurting for cash. But it was good to see that Mark had stood up for himself.

Eduardo calmed down a bit, placing Mark's letter back on the stack of computer books, along with the cease-and-desist order. If Mark wasn't scared, he wasn't going to be either; after all, he hadn't met with the twins, he wasn't a computer coder, and he could only go by what Mark had told him about the differences between the two Web sites. The way Mark painted it, it was like one furniture maker trying to sue someone for designing a new kind of chair. There were thousands of different types of chairs, and making one didn't give you the right to own them all.

Maybe it was a simplified way of looking at the issue—but fuck it, they were college kids, not lawyers. The last thing they wanted to do was get into some bullshit legal battle. Over a Web site that was, perhaps, about to get them both laid.

"Their names were Kelly and Alice—" Eduardo started, but before he could finish, the door to the dorm room opened, nearly hitting Eduardo in the back. Eduardo turned to see Mark's two roommates enter, as disparate-looking a pair of college kids anyone could imagine.

Dustin Moskovitz, in front, was baby-faced and dark-haired, with thick eyebrows and a very determined look in his equally dark eyes. He was quiet, kind of withdrawn, an economics major and a whiz with computers, also incredibly affable, a truly nice guy. Chris Hughes was the far more flamboyant of the two; shaggy blond hair, extroverted, outspoken, with traces of a Southern accent from his upbringing in Hickory, North Carolina. In high school, Chris had been president of the Young Democrats Society and could easily be described as an activist on a number of liberal issues. A bit of a fashionista, he gave Eduardo a run for the most presentable of the group; though where Eduardo chose conservative blazers and ties, Chris favored designer shirts and pants. Sometimes, Mark called him "Prada" because of the way he looked.

The four of them, together—Mark, Eduardo, Dustin, and Chris—were certainly not what you'd call part of the social elite at Harvard. In fact, they'd probably be outsiders at any college, not just the home to Rockefellers and Roosevelts. They were all geeks, each in his own way. But they'd found one another—and something else.

Mark started the conversation, because it was something he'd already decided—and Eduardo was rapidly realizing that that's the way things worked, in Mark's world. Thefacebook was growing fast, and Mark was having trouble keeping up with it all. He was in real danger of flunking some of his classes—and if he wanted to keep thefacebook growing, he was going to need help.

Dustin could handle the computer stuff that Mark couldn't do himself. And Chris was a talker—better than any of them, that was for sure—so he could take charge of the publicity and outreach. The *Crimson* had been a great friend so far; it turned out, Mark had done some IT work for the student paper during his freshman year, which explained all the glowing articles. But going forward, they were going to need to keep on top of the press, because so much of Facebook was about getting people excited, interested enough to log on.

Eduardo would still handle the business side of things—if, indeed, there would be a business side of things. The four of them would be the team to take Facebook to the next level. And they were all going to have titles. Eduardo was going to be CFO. Dustin, vice president and head of programming. Chris, director of publicity. And Mark—founder, master and commander, and enemy of the state. Mark's words, Mark's sense of humor.

Eduardo listened to it all, contemplating what it meant. He knew that things had been much simpler when it was just him and Mark; but he also knew that running a company meant employees, and they didn't exactly have revenues coming in to pay for people's help. So the only option was adding more partners. Mark's roommates were smart, and trustworthy. They were geeks, just like him. And this was a dorm-room operation anyway.

He agreed to the new leadership, and he also agreed to restructuring their ownership agreement. Dustin would get around 5 percent of the company, Chris would get a percentage that would be more fully worked out later on, when they figured out how much work he'd be doing. Mark would drop his ownership down to 65 percent. And Eduardo would own 30 percent. It seemed more than fair. And anyway, there wasn't any money coming in yet, so why haggle over 30 percent of nothing?

"First order of business," Mark said, when that was settled. "I think it's time we open thefacebook up to other schools. Expansion seems like the natural thing."

They'd conquered Harvard, it was time to see how much farther their model could go. They agreed to begin with just a few other elite schools. Yale, Columbia, and Stanford, to start. The site would stay exclusive—you'd have to have an e-mail address from one of those schools to join. Eventually, the community could get bigger, and they'd allow cross-college pollination. Facebook *had* to keep getting bigger.

"But we also have to start talking to advertisers," Eduardo chimed in, refusing to let the issue go. "We need to start monetizing this."

Mark nodded, but Eduardo was pretty sure he didn't entirely agree. Mark knew they should try to make enough money to offset the server costs—but he didn't seem to care about money beyond what it took to run the site. Eduardo felt differently.

Eduardo was starting to believe, in his heart, that they were going to get rich from this Web site. As he looked around the room, at the team of über-geeks they'd assembled—it seemed like nothing could stand in their way.

▶ ▶ ▶

Four hours later, Eduardo's heart slammed in his chest as he careened forward into the bathroom stall, his Italian leather shoes skidding against the tiled linoleum floor. The tall, slender Asian girl was straddling him, her long, bare legs wrapped around his waist, her skirt riding upward, her lithe body arching, as he pressed her back against the stall. His hands roamed under her open white shirt, tracing the soft material of her red bra, his fingers lingering over her perky, round breasts, touching the silky texture of her perfect caramel skin. She gasped, her lips closing against the side of his neck, her tongue leaping out, tasting him. His entire body started to quiver, and he rocked forward, pushing her harder against the stall, feeling her writhe into him. His lips found her ear and she gasped again—

And then another sound reverberated through the bathroom. Something slamming against the stall's wall from the other side of the cold aluminum—then a curse, followed by laughter. A second later, the laughter stopped, replaced by soft moans, and the sound of lips against lips.

Eduardo grinned; now he and Mark shared more than a Web site, they also shared an experience. The men's room of a dorm building wasn't exactly the stacks at Widener Library, but it was something.

As Eduardo went back to the girl wrapped around his waist, bolstered by the music of his friend getting crazy in the stall next to him, a thought hit him, and he couldn't stop smiling.

They had groupies.

And beyond that, he realized, he had been very wrong about something.

A computer program *could* actually get you laid.

CHAPTER 16 | *VERITAS*

The woman behind the reception desk was trying not to stare. She was pretending to fidget with her Rolodex, her fingers parsing through the switches of laminated paper as her bun of dark hair bobbed up and down, but every now and then Tyler could see it, that quick flick of her pale green eyes. She couldn't help looking at them, sitting next to each other on the uncomfortable couch in the waiting area in front of her desk. Tyler didn't blame her; she looked almost as tired as the building itself, and if he and his identical twin brother could provide a little entertainment for the poor, overworked woman, then it was their good deed for the day. Hell, if he'd have thought it would help them with the task ahead, he and Cameron would have dressed exactly the same, like when they were toddlers; though showing up to the Harvard University president's office in striped pajamas and a beanie might have seemed a little disrespectful. Dark blazers and ties seemed more appropriate, and the outer-office receptionist didn't seem to mind. At least, she couldn't stop looking, no matter how hard she pretended she wasn't. And who still used Rolodexes, anyways?

The truth was, Tyler wasn't going to complain about any form of attention, after the week they'd just had. He was sick and tired of being ignored. First, the senior tutor of Pforzheimer House, who had been sympathetic, but had simply passed their complaints along to the ad board's office. Then the ad-board deans, who'd also seemed sympathetic, had read through their ten page complaint against Zuckerberg—then had decided that for whatever reason, it was beyond their jurisdiction. And Zuckerberg himself—who'd responded to their cease-and-desist letter with a bullshit letter of his own. Zuckerberg maintained that he hadn't started work on his thefacebook.com until after their last meeting on January 15; which seemed odd, considering that he'd registered the domain name thefacebook.com on January 13. Zuckerberg had also maintained that he'd just been trying to help out fellow students—for free, out of his own generosity—and that their site was nothing like his.

The kid's response had gotten such a rise out of Tyler and his partners that they'd tried to contact Mark directly. They'd gone back and forth over e-mail and on the phone a bit, trying to get him to meet with them personally. At one point, he'd agreed to meet—but for some reason, only with Cameron. Then that meeting had fallen through, and all contact had ceased. Which, to Tyler, seemed like a good idea, because he didn't think he could trust Mark anyway. He figured that if, in his opinion, Mark had been willing to lie directly to his face, why would a meeting do any good?

So here they were, sitting next to each other on a couch that felt as old as Massachusetts Hall itself, being gawked at by a receptionist. To Tyler, everything about this place seemed ancient. Indeed, Mass

Hall, built in 1720, was the oldest building in Harvard Yard, and one of the two oldest university buildings in the country. The entrance to the building was perpendicular to University Hall, where the legendary statue of John Harvard stood; the statue was constantly referred to by the school tour guides that always seemed to be shepherding groups of prospective students through the Yard as "the statue of three lies," because the words carved into its base—JOHN HARVARD, FOUNDER, 1638 were actually false—as it wasn't actually a statue depicting John Harvard, John Harvard didn't actually found Harvard, and the college was really founded in 1636. Even so, the statue was often the target of pranks by students from other Ivy schools. Dartmouth kids painted the thing green when their football team was in town; Yalies tried to paint it blue, or stuff some replica of a bulldog on its lap. Every school had its own tradition, and even Harvard kids visited the statue in the middle of the night—to urinate on its feet, supposedly for good luck.

Tyler wondered if he and his brother should have tried the urination ritual before they had headed past the statue and into the stultifying air of Mass Hall. They needed all the good luck they could muster. Simply getting an audience with the president of Harvard had been no easy feat. They'd pulled every connection they could find—family, the Porc, friends of friends. And now that they were sitting there, in the waiting room of the ultimate power on campus—it was hard to fight off a looming sense of dread.

When the phone on the receptionist's desk burst to life, Tyler nearly slid off the couch. The woman grabbed the receiver, nodded, then glanced in their direction.

"The president will see you now."

She pointed a hand at a door to her right. Tyler took a deep breath and followed his brother toward the door. As Cameron reached for the knob, Tyler smiled at the woman, silently begging her to wish them good luck. At least she smiled back.

The president's office was actually smaller than Tyler had expected, but well appointed in true academic fashion. There were bookshelves lining one wall, a huge wooden desk, a bunch of antique-looking side tables, and a small sitting area atop an Oriental carpet. On the desk, Tyler noticed a Dell desktop computer. The Dell was significant, because it was the first computer to ever sit in the president's office; Larry Summers's predecessor, Neil Rudenstine, had hated the devices, refusing to allow any computers in his office. The fact that Summers was technologically savvy was a good sign—at the very least he'd understand the issue at hand.

Apart from the computer, the antique side tables told Tyler everything he needed to know about the president. Next to the obligatory photos of the man's children stood framed, signed photos of Summers with Bill Clinton and Al Gore. Next to that, a framed one-dollar bill—signed by Summers himself, a symbol of his time as treasury secretary of the United States, a position he'd filled from 1999 to 2000. A graduate of MIT, Summers had received a doctoral degree in economics at Harvard, had then become one of the youngest tenured professors in the history of the university—at age twenty-eight. After his stint in Washington, he'd returned to Harvard as the twenty-seventh president of the university. His résumé was impressive, and Tyler knew that if anyone had the power to step in and fix the situation, it was Summers.

As they entered the office, Summers was sitting in a leather chair behind his desk, a phone pressed against his ear. A few feet away sat his executive assistant—a pleasant-looking African American woman, maybe midforties, in a conservative pantsuit combination that went well with the room's decor. She waved them both in, pointing to the chairs in front of the desk.

Without hanging up, Summers watched them until they took their seats. Then he continued talking in a low voice for another few minutes, to whoever was on the other end of the phone. Tyler pictured Bill Clinton, maybe on a plane on his way to a speaking gig. Or Al Gore in a forest somewhere, commiserating with the trees.

Summers finally hung up the phone and looked them over. The president had a pudgy, wide face, thinning hair, and barely any chin; his eyes were like pinpoints, slicing back and forth between Tyler and Cameron.

Slowly, Summers leaned forward, and his chubby hand crawled across his desk. His fingers found a stack of printed pages, and he lifted them up by the corner. Tyler could see, immediately, that it was the ten-page complaint that Cameron and he had typed up, detailing all of the conversations they'd had with Mark Zuckerberg, and the time line of their association, from the very first e-mail that Divya had sent to the day the *Crimson* had published the article on the launch of Facebook. Those ten pages represented a lot of work, and it was heartening to see they had reached all the way to the president's desk.

But then, Summers did something that took Tyler and Cameron completely by surprise. Without a word, he took the pages by the corner, and held them up in front of himself like they were covered in shit. He leaned back in his chair—put his feet up on his desk, and stared at the brothers with pure distaste in his eyes.

"Why are you here?"

Tyler coughed, his face turning red. He glanced at the African American woman, who was dutifully taking notes; she'd already written Summers's question down across the top of a blank sheet of lined notebook paper.

Tyler turned back to the president. The disdain in Summers's voice was palpable. Tyler gestured toward the pages hanging from the man's pudgy fingers. He pointed to the front page, the letter he and Cameron had sent to the president's office, outlining their case:

Letter to Lawrence H. Summers, President of Harvard University

Dear President Summers:

We (Cameron Winklevoss '04, Divya Narendra '04, and Tyler Winklevoss '04) are writing to request an appointment with you. We would like to talk to you about a complaint we recently presented to the AD Board, which they declined to bring forward. Our complaint is a well documented case of a sophomore student who broke the honor code with respect to not being honest and forthright in his dealing with members of the Harvard Community.

"The College expects that all students will be honest and forthcoming in their dealings with members of this community" (Student Handbook).

To give you a brief synopsis, earlier this academic year the three of us approached this student (as we had done with former students) to work on our website project. He agreed to work on our site and this began our three month working relationship with him. Over those three months, in breach of our agreement, and to our material detriment in reliance upon his misrepresentations, this student stalled the development of our website, while he began developing his own website (thefacebook.com) in unfair competition with ours, and without our knowledge or agreement.

We are being led to believe this issue falls outside of the realm of academics and the like; however, we believe this student's actions are in direct violation of the Resolution on Rights and Responsibilities adopted by the Faculty of Arts and Sciences on April 14, 1970, which states the following:

"By accepting membership in the University, an individual joins a community ideally characterized by free expression, free inquiry, intellectual honest, respect for the dignity of others, and openness to constructive change."

As the leader of our University, we think you should be aware of incidents that abuse the honor code and threaten the standards of the community. We believe the ramifications of Harvard not addressing this issue will have long-term negative effects throughout the school community and beyond. Therefore, we are requesting a meeting to speak with you about this matter at your earliest convenience. Thank you.

Sincerely,
Cameron Winklevoss '04
Divya Narendra '04
Tyler Winklevoss '04

After he'd let a few seconds pass, so the man could at least pretend to reread their letter, Tyler cleared his throat.

"I think it's pretty self-explanatory. Mark stole our idea."

"So what do you want me to do about it?"

Tyler stared at the man in shock. He turned, and looked at his brother. Cameron seemed just as flabbergasted, his mouth hanging open as he watched the pages sway in the president's pincerlike grip.

Tyler blinked, hard, letting the anger inside him push away the shock. He pointed toward the bookshelf behind the president, where he could clearly see a row of Harvard Handbooks from years past. The handbook was given out to every freshman; inside, it listed all of the rules of the university, all of the codes the administration was supposed to uphold.

"It's against university rules to steal from another student," Tyler said, then added a quote from the handbook from memory: " 'The College expects that all students will be honest and forthcoming in their dealing with the members of this community. All students are required to respect private and public ownership; instances of theft, misappropriation, or unauthorized use of or damage to property or materials will result in disciplinary action, including the requirement to withdraw from the college.' If Mark had gone into our dorm room and taken our computer, you would kick him out of school. Well, he's done something much worse. He's taken our idea, and our work, and the university should step in and uphold the Harvard code of ethics."

Summers sighed, letting the ten pages flop back onto his desk. Tyler watched as they landed next to a pile of brightly colored juggler's balls. Rumor was, the balls had been given to the president by his predecessor, because that's what a president did—juggled things, people, projects, problems. Tyler could tell, from the look on Summers's face, that he and his brother were about to get juggled right out of the room.

"I've read your complaint. And I've read Mark's response. I don't see this as a university issue."

"But there's a code of ethics," Cameron interrupted, forgetting, for a moment, that this was the president, seeing only a pudgy, disdainful man shitting all over the hard work they'd done. "There's an honor code. What good is a code if it doesn't have any teeth?"

Summers shook his head. His jowls reverberated with the motion, like fleshy waves in a swirling epidermal storm.

"You entered into a code of ethics with the university—not with each other. This issue is between you guys and Mark Zuckerberg."

Tyler felt himself sinking into his seat. He felt . . . betrayed. By this man, by the system, by the university itself. He had always seen himself as a member of the Harvard community, as part of an honorable, ordered world. Now the titular head of that world was telling him that there was no community—that it was every geek for himself. Mark had hacked the system, but it wasn't Summers's problem.

"But the university has a responsibility to uphold the honor code—"

"The university isn't equipped to handle a situation like this. This is a technical dispute between students."

"What do you propose we do about it?" Tyler asked, defeated.

Summers shrugged. His rounded shoulders were like two trapped creatures beneath the material of his shirt. It was obvious from his silence that he really didn't care what Tyler and Cameron did about the situation.

"Work it out with him. Or find some other way to deal with it, as a legal issue."

Tyler understood what the man was implying. A face-to-face talk with Mark—which would get them nowhere, considering that the kid was full well willing to lie to their faces. Or a lawsuit. Which seemed even more horrible an option.

It was truly depressing. The president of the university was telling them that they were on their own. The administration was washing its hands of the whole thing. Thefacebook was a popular campus phenomenon. Mark was getting famous, his Web site was growing daily—and the president was basically endorsing his success.

Maybe Summers sincerely didn't think the Winklevoss twins had a case against the kid. Maybe he believed what Mark had written—that the sites were too different, that the Winklevosses were just angry they hadn't been able to launch their project first. Or maybe he just didn't care.

Tyler rose from his chair as Summers basically waved them away.

The only thing left, Tyler realized, was to go after Mark themselves. As he led his brother out of the president's office, Tyler glanced back, watching as the pudgy man went back to his phone. Tyler knew he would remember this moment, because he felt very strongly that it was the true end of his innocence.

To Tyler Winklevoss—whether wrong or right—that damn kid had stolen his idea and made it his own.

And if Harvard had its way, Mark Zuckerberg was going to get away with it.

CHAPTER 17 | MARCH 2004

What a long strange trip it's been . . .

It isn't difficult to imagine the details of that morning sometime in March of 2004, even though the moment itself became historical only in retrospect: Sean Parker's eyes flashed open as he came awake to that sudden and musical thought bouncing around in his brain, a frenetic little ear worm let loose through the thin membrane of his aural canal, infecting his gray matter, powering up the synapses, flicking all the red lights to green. He grinned, as he usually did in the morning, staring up at a blank white ceiling, trying to remember where he was. *What a long strange trip it's been.* He rubbed the last gasps of sleep out of his eyes, then stretched his arms up above his head, feeling the cool, plush material of the heavy down pillow—and it all came back to him.

He was lying on a bed pushed right up against a blandly colored wall of a little bedroom, his head sunk deep into that pillow. His hair was a mess, a tangle of brown-blond curls mushrooming out against the soft material of the pillowcase.

He was wearing a T-shirt and sweatpants, but that was only because it was six in the morning; his Armani jacket, skinny-legged pitch-black DKNY jeans, and tailored Prada shirt were hanging from a hook on the back of the door to the bathroom.

What a long strange trip it's been. His grin turned Cheshire, stretching out the edges of his lips so far it almost hurt. Yes, he knew exactly where he was—and it was a fucking awesome place to be.

He looked around his little bedroom, taking in the little wooden dresser, the bookshelf full of computer textbooks, the lamp in the corner, the sleeping laptop on the miniature side table by the bed. There were clothes strewn all over the place, on the floor, the bookshelf, even hanging from the lamp, but Sean didn't mind because most of them were his clothes, and the ones that weren't were pretty damn sexy. He saw a frilly bra and too short skirt, a tank top and tight, stylish belt—the kind of clothes that college girls wore on campuses all over California; even here, up north, where the palm trees were more often draped in fog than in sunlight. Thankfully, at Stanford, girls still dressed California, despite the school's elite status. And of course, they were all blond. Let the angry brunettes have the Ivies, blond and pretty ruled the West.

Sean pushed himself up on one elbow. He wasn't sure whose bra, skirt, tank top, and belt were in his room—he assumed it was either a guest of one of his roommates, or someone who had been there visiting him. He wasn't certain why the clothes were in his room, either. He might have known the girl, he might not have. Either way, she probably knew him—or at least, she thought she did. It seemed like everybody at Stanford knew Sean Parker. Which was kind of funny, considering that he wasn't a student there. This house that he was living in was full of Stanford kids—it was really just an extension of the dorms, right next to campus. But Sean wasn't a

Stanford student; he hadn't even gone to college. But he was still a campus hero.

Not quite as famous as his original business partner—Shawn Fanning—but those who knew the story, knew the story. The two teenagers who'd changed the record industry by creating a file-sharing Web site called Napster—a site that let college kids everywhere get whatever music they wanted for free, in the privacy of their dorm rooms, by sharing with one another over the Internet. Napster was a massive success, a world-changing creation—well, okay, it had also kind of imploded—but it had been a beautiful implosion.

Napster—which Sean had cofounded after meeting Fanning in an Internet chat room while they were both still in high school—was less a company than a revolution. Napster had made music free, had made it downloadable—had given every kid with a computer real power to get what they wanted. Freedom—wasn't that what rock and roll had been all about? Wasn't that what the Internet was supposed to be about?

Of course, the record companies hadn't seen it that way. The fucking record companies had descended on the two Seans like Vengeful Harpies. They'd battled back, but the end was really a foregone conclusion. Some people thought it was Sean Parker's fault, when it all finally tumbled and fell; according to some printed reports, he'd written some e-mails that had ended up helping out the record companies in their legal battle, a foolish, youthful indiscretion that had cost Napster the endgame—but see, that had always been Sean's problem, and also his strength. He was out there, he didn't keep anything inside.

And he didn't regret anything. No fucking way, that wasn't his style.

Sure, he could have curled up into a ball after Napster had collapsed. Or run home to his parents. But instead, he'd gotten right back on that Silicon horse. Just a couple short years later, he and two of his closest friends had come up with an idea that built on the notion of sharing—but this time, they'd focused on e-mails and contact information. It started as a free system, just a little program that would send out requests for updated info—and it turned into a sort of constant, self-renovating online business card system. They'd called the company Plaxo.

And then, well, in Sean's view that had kind of imploded as well. Not the company—Plaxo was still doing great, the business was probably now worth millions—but Sean's participation in it was over, finished, kaput. In his view, he'd been kicked out of his own company—and it had been even uglier than it sounded.

Ugly, because in Sean's mind, there had been a real villain involved—a James Bond kind of villain, a bizarre, secretive Welshman with a megalomaniacal streak almost as big as his bank account. It had been Sean's idea to bring in the VC monster in the beginning—because he'd thought that Plaxo needed the money, and he'd thought that he knew how to deal with VCs. But Michael Moritz wasn't just any VC, he was one of the partners at Sequoia Capital and a deity among the Silicon Valley moneymen. He'd invested in both Yahoo and Google, made such a fortune that nobody would ever question his methods again.

In Sean's view, Moritz was reclusive, mysterious, and also maniacal. From the start, he and Sean were butting heads on almost every issue. Sean was a freethinker, a young and wild entrepreneur; Moritz seemed to be about money, pure and simple. Barely a year after Seqouia funded the company, Sean believed that Moritz decided that Sean had to go—leave the company he'd founded!—and of course he'd refused. It became a pitched battle, a VC coup—and eventually,

Sean had begun to realize that he was going to end up on the losing end of the situation. His two closest friends, whom he'd started the company with—in Sean's eyes, they'd succumbed to the pressure of Moritz and the board; and according to reported accounts, when Sean tried fighting back by saying that the only way he'd leave was if he could sell a chunk of his ownership in the company for money up front—it pushed Sequoia into war mode. Sean believed that Moritz had done the kind of thing that one would expect a James Bond villain to do; Sean was certain he'd hired a private eye to follow Sean around, to try to get the ammunition necessary to force him to leave.

Sean had started to notice cars with dark windows following him when he left his apartment. He'd noticed strange clicks when he was on the phone, and even bizarre callbacks on his cell phone, from unlisted numbers. It had started to get terrifying.

And maybe they really had been getting dirt. Like any kid his age—with the fame he'd acquired through Napster and Plaxo—Sean liked to party. He liked girls. He certainly wasn't a saint. He was in his early twenties, a kind of Silicon Valley rock star; and he talked really fast, thought really fast. There was a certain jerky, frenetic quality to him—a quality that could be easily misinterpreted.

So maybe they had something on him—maybe they didn't. In any event, in Sean's view Moritz locked him out. Made him resign from his own company. Made him hand over the keys to his own fucking creation. At the same time, Sean believed he had lost both a company and his two former best friends. It had been ugly, and it had been pathetic, and in Sean's view it had been unfair. But, well, it had happened. Not just to him—in Silicon Valley, it happened all the time.

That was the thing about VC money. It was awesome—until it wasn't.

Plaxo had ended badly, but that hadn't meant it was over for Sean Parker. Not even close. The Silicon Valley gossip rags had gotten even more excited about him after the twofer of Napster and Plaxo, and they began to paint him as this bad boy around town. The girls. The designer clothes. And of course, unsubstantiated stories about drugs. Coke. Pills. God knew what else. Sean was half expecting to open up Gawker one day and read about himself mainlining baby seal blood.

The idea that he was a bad boy was kind of funny to him. He guessed it was utterly hilarious to anyone who'd known him growing up in Chantilly, Virginia. He was a skinny kid, allergic to peanuts, bees, and shellfish, and carried an EpiPen filled with adrenaline with him wherever he went. He had asthma, and also carried an inhaler. He had hair that was so unruly it sometimes veered toward an Afro. And okay, skinny was kind of an understatement; he wasn't exactly intimidating, physically. The twin bed was big enough for him to do a gymnastics floor routine. Bad boy of Silicon Valley? The idea was almost ludicrous.

He looked at the frilly bra on the floor of his room, and smiled again.

Okay, maybe he did have his moments. A slight hedonistic streak. As the private eyes probably discovered, he liked girls. Sometimes lots of girls. He liked to go out late and he liked to drink. He'd been kicked out of a few nightclubs. And, well, he hadn't gone to college. He'd left high school when Napster took off and hadn't looked back.

But he wasn't a bad guy. He was the good guy. In his view, even a superhero, kind of. Although his last name was Parker, he thought of himself more as a Batman. Bruce Wayne during the day, hanging with the CEOs and the entrepreneurs. The Caped Crusader at night, trying to change the world one liberated college kid at a time.

Except, unlike Bruce Wayne, Sean didn't have any money yet. He had created two of the biggest Internet companies in history, and he didn't have a dime. Sure, Plaxo was going to be worth something, someday. He'd get a big chunk of that, maybe even tens of millions. Maybe hundreds of millions. And Napster, if it hadn't made him rich, had certainly put him on the map. Some people even already compared him to Jim Clark, the founder of Silicon Graphics, who had been responsible for both Netscape and Healtheon. Sean had already hit two of them out of the ballpark; he only needed a third to make the analogy fair.

And in that regard, he was constantly on the lookout for his next home run. This time, he was looking for something really life changing. Sure, everyone was looking for the next big thing. The difference was, Sean *knew* what the next big thing was. He knew with a complete, and almost religious certainty:

Social Networks.

Just a few months ago, he'd made some connections at the social network site Friendster. He'd brought them some series D VC funding, introduced them to his buddies around town—most notably, Peter Thiel, the guy behind PayPal, a colleague who'd also experienced some run-ins with the gang at Sequoia.

But Friendster wasn't going to be Sean Parker's next home run; it was already too far along, and Sean wasn't getting in anywhere near the ground floor. And to be honest, Friendster had its limitations. It was really a dating Web site. A good one, more disguised than Match or JDate, but it was about meeting chicks you didn't know and trying to get their e-mail.

Then there was MySpace, the ascendant fledgling site that was growing real fast, which Sean had also looked into, and decided against. MySpace was great for what it was, but to Sean, it wasn't really a social network. You didn't go on MySpace to communicate, you went there to show yourself off. It was like one big narcissistic playground. Look at me! Look at me! Look at my Garage Band, Comedy Routine, Acting Reel, Modeling Portfolio, and on and on and on. It was throwing your brand out there and hoping someone paid attention to you.

So if Friendster was a dating sight and MySpace a branding tool, what did that leave? Sean wasn't sure—but somewhere, out there, he knew there was a Fanning plugging away in some basement, working on the Napster of social networking. Sean just had to keep his eyes open.

He knew he had set the bar really fucking high. If it wasn't a billion-dollar company—his own YouTube, his Google—then it wasn't worth his time. But he'd already had a Plaxo, and the experience had been less than satisfying.

The next time it would be a billion dollars or bust.

Sean pushed himself to a sitting position, the energy rising inside of him. It was time to get back to his quest. He glanced at the small table next to the futon, noticing the open laptop resting next to a pink girl's watch. It wasn't his laptop, so it was either one of his roommates or one of his or their houseguests'; either way, it was close enough that he could reach it from bed, which made it the default first choice. It was time to check his e-mails, and begin his morning routine.

He reached for the laptop and placed it gently on his lap. A few seconds later, the computer came out of sleep mode. He saw immediately that it was already hooked up to the Internet, through the Stanford network. He also noticed that there was a Web site open across the screen. Obviously, whoever owned the laptop had been online the night before. Curious, Sean scrolled down, checking the site out.

It was something Sean had never seen before. Which was weird, because he'd seen pretty much everything.

There was a soft blue band across the top and bottom of the site. It was obviously a portal of some sort. A girl's picture was on the left side—Sean took in her beautiful blond hair, her wonderful smile, her incredible blue eyes. Then he saw that beneath her picture, there was some info about her.

Her sex: female. That she was single. That she was interested in boys. That she was looking for friends. And then a list of the friends that she already had found, her networks. The books she liked. The courses she was taking at Stanford.

Next to her profile was a personal quote she'd written herself, as well as some comments from her classmates. Everyone seemed to be from Stanford, with Stanford e-mails. They were her real friends, her actual friends—not people just trying to fuck her, like with Friendster. Not people just trying to show off their new rock band or their new fashion line, like MySpace. This was her actual social network, online, connected. Continually connected. Even when the computer had been sleeping, the social network had been awake. It wasn't static.

It was fluid.

It was simple.

It was beautiful.

"Mother of God," Sean murmured to himself.

It was brilliant. He blinked, hard. A social network—aimed at the college market. It seemed so utterly obvious. The one big gap in the social networking market was college—and college was such a perfect market for a social network. College kids were so incredibly social. You had more friends in college than at any other point in your life. MySpace and Friendster missed the one group of people that had the most use for a social network—but this site? This site seemed to take aim straight at the mother lode.

Sean's gaze drifted down to the bottom of the page. There was an odd little line of text.

A Mark Zuckerberg Production.

Sean smiled. Oh, he liked that. He liked that a lot. Whoever had made this site had put his name right on the bottom of the page.

Sean hit some keys, moved over to Google. He started to do a search. To his surprise, he found a lot, much of it culled from a single source—the *Harvard Crimson,* Harvard university's school newspaper.

The Web site was called thefacebook, and had been started by a sophomore about six to eight weeks earlier. In four days, most of the Harvard campus had signed up. By the second week, there had been nearly five thousand members. Then they had opened it up to some other schools. Now it was estimated there were close to fifty thousand members. Stanford, Columbia, Yale—

Christ. This thing was happening fast.

Sean started mumbling to himself. "Thefacebook." Why not just "facebook"? That was the kind of thing that would drive Sean crazy. His mind was always doing that, instinctively cleaning things up, smoothing them out. He realized with a start that even as he was thinking it, his fingers were rubbing back and forth against the futon's sheets, smoothing out the wrinkles. He grinned at himself. Add OCD to the list of neuroses. Get Valleywag on the phone: *bad boy, asthmatic, peanut-allergied, obsessive-compulsive Sean Parker is chasing after a new project . . .*

Because that's exactly what he was going to do. He was going to find this Mark Zuckerberg, and he was going to see how good this kid really was. And if things were as beautiful as they seemed, he was going to help this kid turn Facebook into something huge.

Billion-dollar valuation or bust. Pure and simple. Nothing less could be considered a success.

Sean had already gone two for two, Napster and Plaxo.

Could Facebook be his number three?

CHAPTER 18 | NEW YORK CITY

"Come on, Eduardo. Do you think they're really going to card us? Here?"

The girl was rolling her eyes, and that just made it even worse; Eduardo glared at her, but she had already turned back to the cocktail list, and now Mark was scanning the damn thing, too. Maybe Kelly was right, and nobody was going to ask for their ID. But that was beside the point. Neither she nor Mark was taking this seriously, and it was driving Eduardo crazy. And it wasn't just the restaurant. The whole trip to New York, Mark had been goofing around, pretending this was all just some big joke. Maybe Kelly could get away with it; she was at the dinner only because she happened to be visiting her family in Queens. But Mark was supposed to be in New York on business.

Though they were staying with friends instead of a hotel, Eduardo had picked up the travel and all the food and taxi bills. More accurately, they were paying for it out of thefacebook's bankroll, the quickly dwindling thousand dollars that Eduardo had put in back in January, three and a half months ago. That defined the trip as a business expense—so Mark should have been treating the excursion as serious business.

But he'd done nothing of the sort. For his part, Eduardo had managed to set up a handful of meetings with potential advertisers; none of the meetings had gone particularly well, however, and it hadn't helped that Mark had slept through about half of them—and had spent the other half sitting silently while Eduardo tried to pick up all the slack. Though everyone they'd met had seemed impressed by the number of people they'd gotten to sign up to thefacebook—over seventy-five thousand at last count—nobody was willing to put any significant money into placing ads on the network. They just didn't get it, yet, and advertising on the Internet, in general, was such a dicey thing. It was simply hard to get the advertisers to understand how different thefacebook was. The fact that people who went on thefacebook tended to stay online longer than on almost any other site was lost on them. The even more impressive statistic, that most kids who tried out thefacebook once tended to come back—67 percent every day—was completely beyond their comprehension.

But maybe if Mark had taken it all a little more seriously, things would have gone a bit better. Case in point; here they were, at one of the fanciest new restaurants in New York, and he was sitting there in that damn fleece hoody, his flip-flops bouncing off each other under the table. Granted, they weren't at 66 to meet with a potential advertiser, but it was still business, and Mark should have looked the part. At the very least, he should have tried to look hip, because in this place he stuck out like a sore thumb.

Located on the first floor of the Textile Building in Tribeca, 66 was Jean Georges's newest hot spot, and quite possibly the nicest Chinese restaurant Eduardo had ever seen. Sleek and minimalist, the place was extremely modern, from the twelve-foot-tall curved glass wall that took up much of the entrance to the huge fish tank that separated the dining area from the kitchen. The floor was bamboo, and frosted-glass panels separated the various leather seating areas. There was also a huge, forty-person communal table, next to another frosted wall behind which the bartenders scampered about, their silhouettes dancing back and forth. Chinese red silk banners hung from the ceiling, but otherwise it seemed more fusion than Asian, at least to Eduardo's palate. Since their guest was running late, they'd already ordered some things from the menu: lacquered pork with a shallot-and-ginger confit. Tuna tartar. A lobster claw steamed with ginger and wine. And foie gras jammed into oversize shrimp dumplings. Eduardo's girlfriend hadn't been too thrilled with the items, and he could tell she was just biding her time until they could order dessert—homemade ice cream that came in little tiny Chinese take-out containers. Though if she could convince one of the waiters to give them drinks without checking their ages, she'd forget all about the ice cream.

She probably wasn't a keeper, but Kelly was still tall and pretty, and Eduardo had managed to keep her interested since their episode in the dorm bathroom. Mark had long lost her friend Alice, but no matter, Mark didn't seem to care one way or the other. At the moment, though, Kelly wasn't the biggest issue dominating Eduardo's thoughts. He was much more concerned about the reason they were at the restaurant in the first place—and the guy they were there to meet

Eduardo didn't know much about Sean Parker—but what he'd found out by a simple search on the Internet, he didn't like. Parker was a Silicon Valley animal, a serial entrepreneur who'd crashed out of two of the biggest Internet companies in what sounded like pretty spectacular fashion. To Eduardo, he seemed like some sort of wild man, maybe even a little dangerous. Eduardo had no idea why the guy wanted to talk to them, or what Parker wanted from them. But he was pretty sure he didn't want anything from Parker.

Speak of the devil; Eduardo caught site of Parker first as he stepped out from behind the curved glass entrance. Although it would have been hard to miss the guy—because he was making quite an entrance, bouncing off the walls like some sort of animated cartoon creature, a Tasmanian Devil spinning through the restaurant. He seemed to know everyone as he moved through the place. First, he was saying hi to the hostess while hugging one of the waitresses. Then he was stopping at a nearby table to shake hands with a guy in a suit, while ruffling the hair of the guy's kid, like they were family friends. Christ, who the hell was this character?

He reached their table and smiled; there was a bit of wolf in that grin.

"Sean Parker. You must be Eduardo, and Kelly. And of course, Mark."

Sean reached across the table, going right for Mark—and Eduardo saw it, then and there—the look on Mark's face, the sudden flush in his cheeks and the brightness in his eyes. Pure idol worship. In Eduardo's eyes, to Mark, Sean Parker was a god.

Eduardo should have realized it earlier. Napster was the ultimate geek banner, a battle that had been fought by hackers on the biggest stage of all. Ultimately, the hackers had lost, but that didn't matter, in a way it was still the biggest hack in history. And Sean Parker had survived that, gone on to Plaxo, made a name for himself a second time. Eduardo didn't have to remember what he'd read on Google, because Sean launched right into it himself, after taking a seat next to Kelly and ordering them all drinks from one of the passing waitresses—a friend, of course, from a previous visit.

Sean spun story after story, his energy level beyond incredible.

About Napster, the battles he had fought. About Plaxo, and the even uglier battles he'd barely survived. He was completely open about everything. Life in Silicon Valley. Parties at Stanford and down in L.A. Friends who had become billionaires, and others who were still searching for that big hit. He painted a really exciting picture of his world—and, Eduardo could see, Mark was eating it all up. He looked like he was about to run out of the restaurant and book a plane ticket straight to California.

When Sean finally reached the last of his stories—for the moment, Eduardo assumed—he turned it around, asking them about their most recent progress with thefacebook.

Eduardo started to explain that they were now in twenty-nine schools—but Sean turned right back to Mark, asking him about the strategies they were applying to get the different schools to join up.

Eduardo sat there, a little miffed, as Mark stiltedly explained their strategy by way of an example. He told the Baylor story—how the little Texan University had at first refused to adopt thefacebook, because the school had a social network of its own. So instead of attacking Baylor head-on, they'd made a list of all the schools within a hundred-mile radius of it, and had dropped thefacebook into those schools first. Pretty soon all the kids at Baylor were seeing all their friends on the Web site—and they practically begged for thefacebook on their campus. Within days, the Baylor social Web site was history.

Sean seemed really excited by the story. He then added to it, by quoting something he'd read in the Stanford newspaper—the *Stanford Daily*—on March 5: "Classes are being skipped. Work is being ignored. Students are spending hours in front of the computer in utter fascination. The facebook.com craze has swept through campus." After that article had come out, 85 percent of Stanford had joined thefacebook within twenty-four hours.

Mark seemed thrilled that Sean had been reading up on him. And Sean, for his part, seemed happy that Mark was a fan. They had an instant connection, there was no denying it. As for Eduardo—well, Sean wasn't purposely ignoring Eduardo, but he was definitely paying a lot more attention to Mark. Maybe it was just the fact that they were both computer savvy—but then again, Sean didn't strike Mark as a computer geek. He was a geek, sure, but his geekiness seemed more chic, like he was just playing a geek on some prime-time television show. It wasn't just the way he was dressed or his amped-up demeanor. It was the way he handled the room, not just their table. He was a showman, and he was damn good at what he did.

The dinner went pretty fast, after that—although it seemed like forever to Eduardo, who almost applauded when Kelly finally got her ice cream. Once the Chinese take-out boxes were all empty, Sean picked up the check, excused himself, and promised Mark that they'd talk again soon. Then the whirling dervish was gone, as quickly as he'd appeared.

▸ ▸ ▸

Ten minutes later, Eduardo was standing next to Mark on the sidewalk outside the restaurant, his hand in the air as he tried to hail a cab. Eduardo's girl had gone off to meet Sean and his girlfriend, to some bar nearby in Tribeca where they were meeting mutual friends. Eduardo was going to meet up with them later, but he still had a few phone calls to make. More advertiser meetings they were trying to set up. He wasn't going to give up, no matter how difficult things got.

Hand still in the air, Eduardo glanced over at Mark. He could see that his friend still had that flushed look on his face. Parker was gone, but his aura still lingered in the air.

"He's like a snake-oil salesman," Eduardo said, trying to break the spell. "I mean, he's a serial entrepreneur. We don't really need him."

Mark shrugged, but didn't respond. Eduardo frowned. He could tell that his words were falling on deaf ears. Mark liked Parker, idolized him. There was no way around it.

Eduardo guessed it didn't really matter, not at the moment. It wasn't like Parker was going to throw money at them; the guy didn't have any real money yet, as far as Eduardo could tell. And thefacebook needed money. As it grew and grew, they were forced to upgrade their servers. And they had also come to the conclusion that they needed to hire a couple more people to work on the programming. Interns, they'd call them, but they'd have to pay them something.

Which was why tomorrow, they were going to open a new bank account, and put some more money into the project. Eduardo had freed up ten thousand dollars to invest into the account. Mark didn't have any funds of his own, so they'd be relying on Eduardo's money for a while longer.

Although Parker didn't have huge funding ability himself, he probably did have some major connections to VC capital. But thankfully—for once—Mark's disinterest in money made that beside the point. For him, the Web site was still primarily about fun, and it had to stay cool. Advertising wasn't cool. VCs weren't cool either. Guys in suits and ties, guys with money—they could never be cool. Eduardo didn't have to worry that Mark would be looking for VC funding anytime soon.

Still, Eduardo couldn't help thinking—to Mark, even despite his VC friends, Sean Parker was the *definition* of cool. But he pushed the thought into the back of his mind. Everything was going so well—he had nothing to worry about. Everyone loved thefacebook.

Sooner or later, they'd figure out how to make money off the damn thing—without the help of Sean Parker. Eduardo had a feeling—Sean Parker couldn't possibly have been the only one who'd taken notice of their little Web site. It was only a matter of time before deep pockets came calling, pockets that could afford a bit more than a dinner at a fancy New York restaurant.

CHAPTER 19 | SPRING SEMESTER

"Yup. It's another one."

"You're shitting me."

"I shit you not."

At first, Eduardo resisted the urge to look over his shoulder. He tried to concentrate on the professor, a bearded, salt-and-pepper-haired man pacing back and forth on the stage at the front of the midsize lecture hall, but it was almost impossible; for one thing, he wasn't even sure what class this was, but it had something to do with an advanced computer language he knew nothing about. Once again, he was crashing one of Mark's lectures. Thefacebook was invading both of their school lives, and even class time was being perverted into makeshift office hours for their burgeoning business. At the moment, the business at hand was fighting that urge not to turn around and stare—which is exactly what he did, because he really couldn't help himself.

It took less than a second to spot the guy—midthirties, gray-suit-and-tie combination, suitcase under his arm—looking completely out of place, sitting between two sophomores in varsity tennis sweatshirts. The guy had a stupid grin on his face—which grew even bigger when he saw Eduardo looking back at him.

Christ. This was getting ridiculous. This wasn't the first VC to track them down on campus; now that the spring semester was almost over and school was getting close to finished, they were coming at an almost frightening frequency. Not just VCs; also reps from the major software and Internet companies. Guys in suits had approached them in the Kirkland dining hall and at the library; one had even found his way to Mark's dorm room, waiting outside for three hours for Mark to come home from a CS department meeting.

The attention was great, but the thing was, they weren't offering real money yet—just the hint that there was money to be had. A few of them had thrown out numbers—nice, big, matzo-ball-type numbers, with seven zeros in them—but nobody had made any real offers, and neither Mark nor Eduardo was inclined to take any of them seriously—even if they had been interested in selling out, which they hadn't even discussed. At the same time, Facebook had now crossed 150,000 members, and was adding thousands more every day. If things continued like that, Eduardo was sure the site was going to be worth serious money. Now that the school year was almost over, he and Mark had to make some important decisions going forward.

Even with Dustin and Chris pulling their weight, thefacebook was beginning to feel like a full-time job. With school ending, it would be easier to balance everything—but thefacebook was certainly going to be a priority for both of them over the summer. Eduardo had made a little progress with advertisers over the past month; he'd been aggressively soliciting on both national and local levels, and had already run free test ads for a handful of big companies—such as AT&T Wireless, America Online, and Monster.com.

He'd also sold some advertisements to a few Harvard undergraduate organizations—the Harvard Bartending Course, the Seneca Club's Red Party, the Mather House's annual "Lather" dance. The College Democrats were paying thirty dollars a day to drum up interest in an upcoming trip to New Hampshire. So the site was earning a little bit of cash. Not quite enough to offset the rapidly growing server costs—and the upgrading and maintenance necessary now that there were so many people on the site, twenty-four hours a day. But it was a start.

Eduardo had also moved the business along in terms of its structure; he and Mark had officially incorporated themselves on April 13, legally creating TheFacebook, LLC, registered in Florida, where Eduardo's family lived. In the incorporation documents, they'd laid out the ownership of the company as they'd agreed upon in Mark's dorm room: 65 percent ownership for Mark, 30 percent for Eduardo, and 5 percent for Dustin. Chris was still going to get some percentage in the future, but that hadn't been decided on yet. In any event, just having those incorporation documents made the company feel more real—even if it wasn't actually making any profits yet.

But even with the incorporation documents, and the continued viral growth of thefacebook, the decision of what to do when school ended in a few weeks was still a difficult one. Both Mark and Eduardo had gone through the motions of looking for summer jobs. Mark hadn't found anything he'd been psyched about, but Eduardo, through his Phoenix connections and his family's friends, had managed to land a pretty prestigious internship at a New York investment bank.

Eduardo had gone back and forth about the internship with his dad—and it had been pretty obvious which way his dad had been leaning. Thefacebook was growing and incredibly popular, but it still wasn't making any real money. The internship was a respectable job, and an amazing opportunity. And since most of the advertisers thefacebook was chasing after were based in New York anyway, didn't it make sense for him to take the internship, and work on thefacebook during his spare time?

Before Eduardo had even been able to bring up the idea with Mark, Mark had dropped a bombshell of his own; although thefacebook was his priority as well, he'd started developing a side project called Wirehog with a couple of his computer programming buddies—Adam D'Angelo, his high school friend with whom he'd invented Synapse, and Andrew McCollum, a classmate and fellow CS major.

Wirehog was basically a bastard child of Napster and Facebook, a sort of file-sharing program with a social network feel. Wirehog would be downloadable software that would allow people to share anything from music to pictures to video with friends, via personalized profile pages linked to other friends in a personally controlled network. The idea was, when Mark was finished with Wirehog, he'd merge it into thefacebook as an application. Meanwhile, both he and Dustin would also be continuing to upgrade thefacebook; they hoped to increase the number of schools using the Web site from about thirty now to over one hundred by the end of the summer.

It was a heady task, especially combined with the Wirehog project. But Mark seemed more thrilled than overwhelmed. And the fact that Mark planned to divide his time between the two projects made Eduardo's decision to take the internship a little easier.

It wasn't until Mark had dropped the second bombshell that Eduardo started to feel a little concerned. Mark had broken the news to Eduardo just yesterday, in fact, after Eduardo had already accepted the internship and had even started looking for rental apartments in New York.

Somewhere in the past few weeks, Mark had explained, in his dorm room over a six-pack of Beck's, he had come to the conclusion that for the next few months, California seemed like the place he should be. He wanted to work on Wirehog and thefacebook in Silicon Valley—a place of legend, to computer programmers like Mark, the land of all of his heroes. Coincidentally, Andrew McCollum had landed a job at Silicon Valley–based EA sports, and Adam D'Angelo was going as well. Mark and his computer friends had even found a cheap sublet on a street called La Jennifer Way in Palo Alto, right near the Stanford campus. To Mark, it seemed like a perfect plan. He'd bring Dustin along, they'd set up shop in the rental house, and thefacebook and Wirehog would be right where they belonged. California. Silicon Valley. The epicenter of the online world.

Even a day later, Eduardo still hadn't come to terms with Mark's second bombshell. In truth, he didn't like the sound of it all; not only was California as far away from New York as you could get—but it was also, to him, a dangerous and seductive place. While Eduardo was off in New York, chasing advertisers, guys in suits like the VC sitting a few rows behind them would be chasing Mark. And even worse than the guys in suits were the guys like Sean Parker—who knew the exact buttons to push. Running the business out of California had never been the plan. Mark and Dustin were supposed to be programmers, while Eduardo was supposed to play the businessman. If they separated, how was Eduardo going to guide the business like they'd agreed?

But Mark had shrugged off Eduardo's concerns when he'd voiced them; there was no reason why they couldn't work from two cities at once. Mark and Dustin would continue programming while Eduardo would find advertisers and handle the finances. In any event, there wasn't time to debate the issue; Mark had already made his decision, and Eduardo had accepted his internship in New York. They'd just have to find a way to make it work.

Eduardo didn't love the idea, but he figured it was only for a few months; then they'd both be back at school, being chased around by VCs in ridiculous gray suits.

"I guess I should go talk to him," Eduardo whispered as he turned away from the man's hundred-watt smile. "You want to come, too? They're always good for a free lunch."

Mark shook his head. "We're interviewing interns today."

Eduardo nodded, remembering. Mark and Dustin had decided that they'd need to bring at least two interns with them to California if they were going to have any chance at reaching a hundred schools by the end of the summer. Which would cost them, of course; nobody was going to follow them across the country for free. The word they'd put out through the CS department was that they were going to pay somewhere in the order of eight thousand dollars for the summer job, along with room and board in the La Jennifer Way sublet. It seemed like a lot—considering that the company wasn't making any money yet—but Eduardo had agreed to fund the project once again, out of his investment earnings. In a few days, he planned to open a new Bank of America account in the company's name. He'd freed up eighteen thousand dollars to deposit into the account, and he was going to give Mark a package of blank checks to fund their operation in California. As the man in charge of the business side of the operation, it seemed the right thing to do.

"After I'm done with this bozo," Eduardo responded, "I'll come by and help out with the interns."

"Should be interesting," Mark responded, and Eduardo was pretty sure he saw the hint of an evil little grin.

Interesting could mean just about anything, in Mark's unusual world.

▶▶▶

"And go!"

We can imagine the scene that Eduardo witnessed when he stepped through the threshold of the basement classroom just as the place exploded; his ears rang from the shouts, raucous laughter, and applause, and he had to push his way through a crowd of onlookers just to see what the hell was going on. The crowd was mostly men, mostly freshmen and sophomores, and all computer programming students—obvious from the pasty pallor of their cheeks to the way they seemed completely comfortable in the low-ceilinged, ultramodern comp lab. They completely ignored Eduardo as he jostled his way to the front of the mob, and when he finally made it through, he could see why. The game was in full swing, and it was infinitely more "interesting" than even he could have imagined.

The center of the computer lab had been cleared out; in the clearing five tables had been lined up next to one another, and on each table sat a laptop computer—next to a row of shot glasses filled with Jack Daniel's whiskey.

Five computer geeks were at the tables, furiously pounding the keyboards of the laptops. At the head of the tables stood Mark, with a timer in his hand.

Eduardo could see the screens from his vantage point—but to him, they were just a jumble of numbers and letters. No doubt the kids at the tables were racing through some byzantine, complex computer code; probably designed by Mark and Dustin to test just how good they really were. When one of the kids reached a point in the code that made the screen blink, he looked up, then downed one of the shots of whiskey. The crowd erupted into applause again, and the kid went right back to his programming.

Eduardo was immediately reminded of the boat race he had taken part in during his initiation into the Phoenix. And this, too, was an initiation of sorts—into Mark's world, the Final Club he had created with his imagination and his computer prowess. It was a race, a test—and probably the oddest interview session for an internship these kids would ever go through; but if it bothered them at all, none of them were showing it. The expressions on their faces were of pure enjoyment. They were hacking while doing shots—proving not only their capability at programming under pressure, but also their willingness to follow Mark anywhere. Not just to California, but wherever he wanted to lead them. To them, Mark wasn't just a classmate. He was rapidly becoming a god.

After ten more minutes of shouting, key slamming, and shot pounding, two of the kids leaped to their feet—almost simultaneously—turning their chairs over behind them.

"We have our winners! Congratulations!"

At that moment, someone hit an MP3 player hooked up to speakers in the corner of the room, and a Dr. Dre song burst out: *California, it's time to party . . .*

Eduardo had to smile. The crowd closed in around him, filling the center space, and then the place was near bedlam, as everyone moved to congratulate the new interns. Eduardo was jostled backward, and he let himself go with the flow, content to just watch Mark have his moment. He saw Mark and Dustin join the interns—forming a little cabal in the center of the room. He also noticed that there was a pretty Asian girl at Mark's side; tall, Chinese, with jet-black hair and a really nice smile. She'd been around Mark a fair amount in the past few weeks. Her name was Priscilla, and he was starting to think that this girl was going to be Mark's girlfriend—a concept that had seemed unthinkable just four months ago.

Things had certainly changed for both of them. For once, Mark looked genuinely happy, in the center of the swarm of idolizing computer programmers. And Eduardo was happy, too, even though he was off to the side, watching.

He decided then and there that they could make it work; he could run the company out of New York while Mark and Dustin, McCollum and the interns did the programming in California. Maybe they'd make some good connections in Silicon Valley while they were there—connections that Eduardo could mine for the better advancement of the site. They were a team, and he would be a team player. Even if that meant watching over them from three thousand miles away.

And anyway, in three months, they'd all be back at school—Eduardo entering his senior year, Mark his junior—and life would continue. Maybe they'd be rich by then. Or maybe they'd be right where they were now, watching their company grow and grow. Either way, they were already far different from when they began this adventure, and Eduardo had no doubt that the future was going to be grand. He pushed any concerns away, because that's what a team player did. There was no need to be paranoid.

Truly, he asked himself, how much could go wrong in a handful of months?

CHAPTER 20 | MAY 2004

"Three."

"Two."

"One . . ."

Tyler felt his fingers whiten against the crystal flute of champagne as he watched Divya and Cameron hunch next to each other over the desktop computer. Divya's finger was in the air, paused over the computer's keyboard; he was drawing this out for all it was worth, trying to make it as dramatic as possible. In theory, the moment *was* dramatic: the launch of the Web site they had worked on since 2002, almost two full years. Renamed ConnectU—mostly to try and help them overcome the trauma of what had gone on over the past few months, but also because now that thefacebook had proven that the idea behind the Harvard Connection could work in many schools simultaneously—the site was finally ready to go online. After so many hours of discussion, planning, anxiety—so many days spent worrying over the design of the site, the graphics, the features. It was a spectacular moment.

And yet, it didn't feel that spectacular—or that dramatic.

Maybe that was because in practice, it was just an Indian kid hitting a key on a computer keyboard while two identical twins watched on from within a stark, almost barren Quad dorm bedroom.

Most of Tyler's belongings had already been packed up in cardboard boxes, which were labeled and stacked around the edges of the small room. His and Cameron's dad would be there in a few hours to help them move out—and then they would be leaving Harvard for good, heading off into the real world. Well, maybe not the real world. Cameron and Tyler were going right into training—an even more intense regimen than they had been following at Harvard. To help them with their mission, their father had revamped a boathouse in Connecticut. They'd hired a coach, and now that they had graduated, they were going to make a serious go at making the Olympics in Beijing in 2008. Between now and then, of course, there would be thousands upon thousands of hours of training. It was going to be hard, painful, and, at times, incredibly aggravating.

But while they trained, ConnectU would be chugging along. Hopefully gaining members in colleges across the country. Hopefully, somehow, competing with thefacebook, MySpace, Friendster, and all the other social networks that were already moving forward, spreading like viruses across the World Wide Web.

Tyler knew they were starting at a huge disadvantage. He knew all about the business concept of "first mover advantage"; his father had taught business at Wharton for twelve years after founding his consulting company, and he'd explained the idea to Tyler many times. For certain industries, it wasn't about quality of product or even corporate strategy. It was about who got there first. It was a landgrab, and ConnectU was coming late to the plains.

Which was exactly what was so damn frustrating about what Mark Zuckerberg had done to them. In Tyler's mind, he hadn't just stolen their idea, he'd also stalled them for two months. If he'd just told them he wasn't going to program their site, they'd have found someone else. They'd have been mad, but they'd have moved forward, and they wouldn't have blamed him for trying to damage their dream. Maybe they'd have launched first—and it would be ConnectU that every college kid in America was talking about. It would be ConnectU that was changing the social lives of so many people.

It was beyond frustrating. Every day, Tyler, Cameron, and Divya had to listen as classmates chatted on and on about thefacebook. And not just at Harvard; the damn thing was everywhere. In the dorm rooms down the hall, on the laptop in every bedroom. On the TV news, almost every week. In the newspapers, sometimes every morning.

Mark Zuckerberg. Mark Zuckerberg. Mark fucking Zuckerberg.

Okay, maybe Tyler was becoming a little obsessed. He knew from Mark's point of view, he, Cameron, and Divya were just a blip in the history of thefacebook. In Mark's mind, he had worked for a few hours for some jocky classmates, gotten bored, and moved on. There were no papers signed, no work agreements or nondisclosures or noncompetes. Mark had bullshit them in e-mails, sure, but in his mind, what did he owe a couple of jocks who couldn't even write computer code? Who were they to try to grasp on now that he was flying so high?

Sure, Tyler had read Mark's letter to the administration, his e-mailed response to Cameron's cease-and-desist. "Originally," Mark had written to Cameron, "I was intrigued by the project and was asked to finish the Connect side of the website. I did this. After this meeting, and not before, I began working on Thefacebook, using none of the same code nor functionality that is present in Harvard

Connection. The only common aspects of the site are that users can upload information about and images of themselves, and that information is searchable."

And he'd also read Mark's more vicious response to the university, when Tyler and Cameron had been trying to get the ad board involved:

> **I try not to get involved with other students' ventures since they are generally too time-consuming and don't provide me with enough room to be creative and do my own thing. I do, however, make an effort to use my skills to help out those who are trying to develop their own ideas for websites. Perhaps there was some confusion, and I can see why they might be upset that I released a successful website while theirs was still unfinished, but I definitely didn't promise them anything. Frankly, I'm kind of appalled that they're threatening me after the work I've done for them free of charge, but after dealing with a bunch of other groups with deep pockets and good legal connections including companies like Microsoft, I can't say I'm surprised.**

But it was the last line of that ad-board letter that really irked Tyler. After trashing their site, Mark had concluded: "I try to shrug it off as a minor annoyance that whenever I do something successful, every capitalist out there wants a piece of the action."

In Tyler's mind, that was utter bullshit. For Tyler, Cameron, and Divya, it wasn't about the money at all. It had never been about money. Tyler didn't give a shit about money. Christ, his family had plenty of money.

It was about honor. It was about fairness. Maybe in business, those things could be pushed to the side. Maybe in a hacker's world, those things took second place to what you could do, how much smarter you were than the other guy. But to Tyler, there was nothing more important than honor.

Obviously, Mark felt differently about the subject. A few times, over the past few weeks, Tyler had thought about just going over to the kid's dorm room and confronting him, face-to-face. But he'd resisted the urge, because he'd known that it wouldn't have gone well.

One night just a week ago, Cameron had, in fact, been coming out of a party at one of the River Houses, when he'd seen Mark standing across the street. When he'd taken a step toward the kid—just to talk—Mark had turned and sprinted away.

There was no doubt in Tyler's mind that the situation would never be resolved by a simple conversation. Things had already gotten too ugly for that. The only choice seemed to be to move forward, as best he could.

As Divya finished his countdown, Tyler shook his angry thoughts away, focusing on his brother and friend in front of the computer. This moment wasn't about Mark Zuckerberg, or thefacebook. This was about ConnectU, and hopefully they were turning a new page in their lives.

"And here we go," Divya continued, his voice rising. "Liftoff!"

His finger came down on the keyboard, the screen blinked—and then it was done. ConnectU had gone live. It was out there, and hopefully, people would notice. Hopefully, college kids would sign on, and the site would grow and grow.

Tyler raised his glass as Divya and Cameron clinked theirs together. Then he took a long drink, feeling the bubbles against his throat. Still, despite the celebratory mood, he couldn't help but notice that the taste in his mouth was exceedingly bitter.

He knew, deep down, that the bitterness had nothing to do with the champagne.

CHAPTER 21 | SERENDIPITY

At its essence, it was simply a matter of physics. Force versus an equal and opposite force. An object in motion tending to stay in motion, no matter how unusual, unwanted, or just plain annoying that motion happened to be. Force equals mass times velocity—there simply wasn't any way around the physics of it; at 150 pounds soaking wet, Sean Parker had no way of stopping the oversize mahogany bureau from caterwauling down the steps of the front porch of the compact little bungalow—so he didn't even try.

Instead, he just stood there shaking his head as the damn thing rolled onto its side, landing with an ugly *thud* in a patch of grass next to the driveway. He waited for a few seconds, listening carefully—but he didn't hear any complaints coming from inside the house, which was a very good thing. Obviously, his girlfriend hadn't heard the thud, which meant that if he could get the now slightly damaged, monstrous piece of furniture into the back of his BMW parked a few yards away in the driveway of the house, she'd never be the wiser.

He bent to one knee, putting his hands underneath the heavy wood, and gave it a solid try. His expensive Italian driving shoes sank a few inches into the grass as his face turned bright red with the effort. He felt his lungs starting to close up a little, and he coughed, quickly giving up. He wondered for a moment if a few hits from his inhaler would make the task any less impossible. Probably not, he decided. More likely, he was going to have to suck it up and ask his girlfriend for help. Not the most manly of options, but then again, he'd been crashing in her pad for much of the last semester of her senior year at Stanford, and now that she was moving back home, it might be nice for them to share one moment of domesticity—even if that moment consisted of lugging a hundred-pound bureau across a tranquil bit of front lawn—

"Sean Parker?"

The voice came out of nowhere, interrupting Sean's silent contemplation of all things bureau-related. He looked up, then realized the voice had come from behind him, down the quiet Palo Alto street where his girlfriend's family lived. He turned on his heels—and squinted, as the sunlight caught him straight in the face.

When his eyes adjusted, he made out four young guys coming toward him. Strange, to see young people in this neighborhood; the sleepy town wasn't exactly the hippest part of the suburban community—a pretty little warren of bungalow-style homes, swimming pools, and manicured lawns, maybe even with the odd palm tree or two—and Sean guessed the average age of the residents was a good thirty years older than these kids looked. College guys, he assumed, from the way they were dressed—sweatshirts, jeans, and one gray hooded fleece between them.

Sean didn't recognize any of the kids at first, but as they got closer, he suddenly realized that he did indeed know one of them.

"This is a bizarre coincidence," he murmured, figuring out who it was.

Mark Zuckerberg seemed as shocked as he was, though it was hard to read the kid's face. Mark quickly introduced his roommates, and explained that they had just recently moved into a house right in the neighborhood—in fact, Mark pointed out the house, which was barely half a block away from Sean's girlfriend's family. Mark and his roomies had literally stumbled on Sean by accident—although Sean had never really believed in accidents like this. Fate, fortune, call it whatever you like, but his whole life had sometimes seemed like a sequence of fortuitous events.

He'd worked so hard to track Mark Zuckerberg down in New York, and now out here in California, the boy genius had stumbled right into his lap. To be sure, since the dinner at 66, he and Mark had made plans a couple of times via e-mail to try to meet up; in fact, only a few weeks earlier they had hoped to coincide in Vegas at some high-tech event, only to have their plans fall through. But this was even better. *Way better.*

As Sean explained his situation—that he was moving his girlfriend into her parents' house now that the semester had concluded, that he was going to be staying with her for a couple of days but after that he would be temporarily homeless—he could see the bright lights going off behind Mark's eyes. After all, Mark had come to Silicon Valley because it seemed like the right place to go to build an Internet company. So what could be better than having an adviser who'd already launched two of the most talked-about companies in town crashing in the same house? Mark didn't make any formal offer, but Sean could tell that the option would be there, if it was something he was interested in—which he knew it would be.

He'd wanted to get involved with thefacebook the minute he'd seen the Web site; if all went well, he was going to be living with the guy who had created it.

You didn't get more involved than that.

▸ ▸ ▸

The kid was flying through the air like Peter Pan in some bizarre, high school production, except instead of being attached to a safety harness and a guide wire, he was hanging on for dear life to a makeshift zip line that had been run from the base of a chimney on the top of the house all the way to a telephone pole on the other side of the swimming pool. The kid was screaming as he went, but Sean could tell he was probably more drunk than scared; still, he managed to launch himself at exactly the right moment, performing an airborne spin that landed him directly in the center of the pool. Water splashed outward, drenching an outdoor barbecue and even reaching the wooden deck that stretched around the back side of the house on La Jennifer Way—that same, quiet suburban street just a few miles outside of Palo Alto's center.

Sean couldn't have been more pleased by the setup; the house was great, with a wonderful frat-house feel to it—even though Mark and his friends had only recently moved into the place. They'd bought the zip line for a hundred dollars at a nearby hardware store, installing it themselves, with only minimal damage—so far—to the chimney or the telephone pole.

The interior of the house hadn't needed much improvement; it had already come furnished, and Mark and his friends had brought little with them. Maybe a bag or two each, and some bedding—and that was all. Mark's parents had sent some fencing equipment, so there were foils and fencing helmets scattered about. They'd also picked up some engineering whiteboards at a local Home Depot—

boards that were already covered with the scrawl of computer code, in numerous bright colors. The floor of the house was littered with empty pizza boxes, beer cans, and the cardboard remains of a fair amount of computer equipment. The oversize living room looked like a mix between a dorm room and an engineering lab—and twenty-four hours a day, there was someone locked into one of the multiple laptops or desktops that were strewn about, wires curling everywhere like the entrails of a downed alien spacecraft. The sound track for the scene was a mix of alternative and hardwired rock—a lot of Green Day, Sean noticed, which seemed appropriate for a group of hacker types with anarchistic streaks.

Sean was likewise happy to see that the team Mark had assembled were perfect engineering soldiers; brilliant, all of them, even the interns—Stephen Dawson-Haggerty, and Erik Shilnick, both freshman CS majors, experts on Linux and front-level coding. Along with Dustin and Andrew McCollum, Mark had the makings of a real brain trust. The work ethic in the house was spectacular; almost literally, the group programmed night and day. Including Mark—especially Mark—when they weren't sleeping, eating, or hurling themselves into the swimming pool via the zip line, they were at the computers. From noon to five in the morning, coding away, adding colleges one after another to thefacebook, working out the kinks, adding applications, and developing Wirehog. They were a top-notch crew, possibly the best start-up raw materials Sean had ever seen.

The one person Sean *didn't* see in the house was Eduardo Saverin. Which, at first, seemed confusing, since back in New York Eduardo had been introduced as the titular business head of thefacebook, and had certainly made it very clear—multiple times—that he was going to be running all the business aspects of the Web site. But it was obvious from the minute Sean walked into the La Jennifer Way house that Eduardo wasn't involved in the day-to-day workings of thefacebook at all.

In fact, Eduardo had gone to New York to pursue some sort of internship at an investment bank, according to Mark. Which immediately set off warning bells in Sean's mind. Having been a part of two major companies—and witnessed many more successes and failures—he knew that the most important aspect of a start-up was the energy and ambition of the founding players. If you were going to do something like this—really do it, really succeed—you had to live and breathe the project. Every minute of every day.

Mark Zuckerberg was living it. He had the drive, the stamina, and the ability. He was obviously a genius—but more than that, he had the strange, unique focus that was necessary to pull something like this off. Watching him program at four, five in the morning—every morning—Sean had no doubt that Mark had the makings of one of the truly great success stories in the modern, revitalized Silicon Valley.

But where was Eduardo Saverin? Or more accurately—was Eduardo Saverin even part of the equation anymore?

Eduardo had seemed like a perfectly nice kid. And of course, he'd been there in the beginning. He'd put up a thousand dollars, according to Mark, to pay for the first servers. And it was his money, at the moment, that was financing the current operation. That gave him some weight, sure, like any investor in a start-up. But beyond that?

Eduardo saw himself as a businessman—but what did that mean, exactly? Silicon Valley wasn't about business—it was an ongoing war. You had to do things out here to survive that weren't taught in any business class. Hell, Sean had never even gone to college, he'd started Napster while still in high school. Bill Gates had never graduated Harvard. None of the true success stories out here had gotten where they were by taking classes. They became successes by coming out here—sometimes with just a duffel bag on their back and a laptop in their hands.

Eduardo wasn't here—and as far as Sean could tell, he wasn't interested in being here. So Sean pretty much put him out of his thoughts. He had Mark, he had Mark's team—he had thefacebook. With his help, he truly believed they could build this company into the billion-dollar project he'd been looking for. Fate had put him in the right place for the third time—hell, he was sleeping on a mattress in an empty corner in the house, most of his belongings still in storage somewhere—and he was going to make this work.

First, he was going to help these guys figure out what it meant to be a part of this revolution—because the way Sean Parker saw it, that's exactly what Silicon Valley was all about. A constant, continuing revolution. He was going to show them this world like only he could.

Looking around this house, at these guys with their fencing equipment and their pizza boxes, he could tell that they could use a little lesson in the finer ways of living this life. After all, they were creating a premier social network. They should at least understand what it meant to be truly social. Sean knew he was just the guy to show them what was possible. He was a rock star in this town—but there was no reason that Mark Zuckerberg couldn't eclipse even him, eventually. Thefacebook was going to be hot—which meant Mark, for all his awkwardness, for all his flaws—he was going to become the toast of the town. Parties, fancy restaurants, girls—Sean could show him the way to all of it.

As for Eduardo, well, it was sad that the kid was going to miss out on the next stage of the company. But that was something that happened all the time in this game. Eduardo had been at the right place, at the right time—but the place had changed, and time was moving forward at the speed of light. Eduardo might try to hang on—but he was already showing that he didn't have what it took.

Poor kid, Sean thought to himself.

What happens when the guy standing next to you catches a lightning bolt? Does it carry you up to the stratosphere along with him?

Or do you simply get charred trying to hold on?

CHAPTER 22 | CALIFORNIA DREAMING

The rain was coming down in fierce gray sheets by the time the American Airlines 757 wide-body began to taxi toward the runway. Eduardo had his face against the circular window, but he couldn't see anything beyond the rain. There was no way to tell how many planes were lined up ahead of them, but since it was JFK, a Friday night, and the weather sucked, there was a good chance they'd be sitting on the runway for a while. Which meant he was going to get into San Francisco well past the ten P.M. expected time of arrival—which would feel like one A.M. to him. He was going to be exhausted by the time Mark and the rest of them picked him up at the airport—but he knew it wasn't going to make any difference. From the sound of the night they had planned, he was going to have to hit the ground running.

The throb of the engines powering up as the plane rolled slowly forward reverberated through his tired muscles, and he settled back against the narrow coach window seat, trying to get comfortable. Even though he was in his customary jacket and tie, he didn't think he was going to have trouble sleeping during the six-hour journey. He had been burning it pretty hard the past month in New York. Ten-hour days spent hitting the pavement, taking meetings with advertisers, potential investors, software makers, anyone who was interested in thefacebook, whatever the reason. Then dinners and nights out in the various New York clubs, mostly with friends from the Phoenix who were also spending the summer in the City; and of course, time spent with Kelly, who was now calling herself his girlfriend, at various times correctly, though he was starting to realize that she was a bit crazy.

He didn't regret—even for a moment—that he had quit his internship on the very first day—really, minutes after he had first sat down in the little cubicle he was supposed to occupy for the next ten weeks, and had stared at that pile of stock valuations he was supposed to analyze—when he'd realized that he wasn't going to become a real businessman like his father by neglecting the business he and Mark had cofounded in the dorms. But he couldn't help but be anxious about thefacebook, especially late at night, wondering how things were going in California with Mark and the rest of the team, what they were up to, what progress they had made—and why they weren't calling more often.

He rolled his eyes at himself as he stretched into the stiff, too-small seat; maybe he was starting to think like the crazy girlfriend he was already considering dumping, maybe being a little jealous. Wasn't that the real reason he had booked the last-minute trip to California, to see for himself that his concerns were unfounded?

By the end of tonight, he was certain things would feel back to normal with thefacebook. He and Mark and the rest would have a blast, get some work done, and everything would be copacetic. And it would all start with a bang.

Mark had said something about a party that Sean Parker had gotten them invited to—some sort of charity bash that all the big-shot entrepreneurs would be attending. It would be fun, but there'd also be the opportunity to meet with more investor types, including some VCs, some major Silicon Valley players, even a few Internet celebs. According to Mark, Parker had already taken them to a handful of similar parties; over the past month since they'd hit California, Mark had seen all the highs the area had to offer. They'd worked their way into the Stanford summer scene, the San Francisco high-tech groove, and had even made a few trips down to L.A. for high-profile Hollywood bashes.

Sean Parker knew everyone, and everyone knew Sean. Through him, everyone was getting to know Mark, too; thefacebook wasn't the biggest kid on the block by any means, but it was slowly becoming the talk of the town, and it seemed like everybody wanted to meet the whiz kid behind the much-hyped social network. Eduardo couldn't help but grow more and more concerned each time he spoke to Mark, and heard about another milestone, party, or dinner that he had missed by being in New York. Worse yet, Mark was Mark—hard enough to read in person, but on the phone he was a complete mystery. Sometimes it was like talking to a computer. He heard what you said, digested it, but responded only if he felt a response was necessary. Sometimes he didn't respond at all.

If he was thrilled that Eduardo had finally made some real progress with advertisers—specifically, landing a deal with Y2M, and getting a few other big players to make some pretty impressive promises—he certainly wasn't showing it. To be fair, Mark and his team were working round the clock at adding features to the site, and signing up more and more schools. At the rate they were going, they would surpass five hundred thousand members by the end of August—a pretty spectacular number. But with that incredible growth, there came new problems.

Most important, they were going to need more money soon. The company was still running off of the eighteen thousand dollars that Eduardo had deposited into the Bank of America account, via the blank checks he'd given Mark when he'd opened the account. The advertising money that was coming in wasn't going to be enough to keep up with the demand; five hundred thousand users would burn a lot of server space. And pretty soon, two interns would not be enough to keep the company running. They'd have to hire real employees, get a real office, hire real lawyers—etc., etc., etc.

All of these things, Eduardo was prepared to discuss—as soon as he could get Mark alone. It wasn't stuff that Parker needed to hear about, because it didn't concern Mark's houseguest, no matter how many parties he took them to.

Eduardo felt a sudden buzzing in his pocket, and he glanced around the plane, momentarily confused. Then he realized with a start that he hadn't turned his cell phone off. He hadn't been getting reception in the taxi over to the airport, but it must have finally found a satellite. He glanced out the window, saw they were still rolling along the tarmac, then yanked the thing out of his pocket.

When he looked at the screen, his lips turned down at the corners.

Twenty-three texts—all from Kelly. Wonderful.

She was in Boston, still in the dorms, taking summer courses. The night before, he had made the foolish mistake of telling her over the phone that he was going to California to hang out with Mark and the boys for a few days. She had immediately reacted badly, voicing all these paranoid ideas that they were going to be partying with girls they'd met on thefacebook. It was a ridiculous notion—although, to be fair, they *had* met a bunch of girls over thefacebook, and more than that, they were becoming pretty well known, on and off campus, because of the Web site. Or at least Mark was—Christ, his name was on every single page.

But Kelly was just being crazy. They weren't going to be partying with random girls, they were going to be working a Silicon Valley scene. Eduardo texted her back, telling her to calm down. He remembered that he'd left her a gift in her dorm-room closet the last time he'd visited—a new jacket, still wrapped up in a gift box from Saks Fifth Avenue. He told her to open it, and that he was thinking about her, and not to worry.

Then he shut off the phone and jammed it back into his pocket. With a thrust of the engines, the plane tipped back, pressing him against the stiff seat. Didn't he have enough to worry about?

The last thing he needed to deal with, right then, was a jealous girlfriend.

▸ ▸ ▸

"Don't be afraid. Okay, be afraid. But it runs pretty well."

Eduardo raised his eyebrows as he followed Mark out of the terminal and caught sight of the car parked right up against the curb; he couldn't even tell what make it was, but it was really old, and the whole thing was trembling. It looked like one of the tires was slightly bigger than the other three, giving the chassis an odd sort of tilt. In other words, the car was really a piece of crap.

Which was exactly as expected, since Mark had bought the thing on Craigslist just a couple of days before. It didn't even use a key, you started it by fidgeting with the ignition. The good thing was, they didn't have to worry about anyone stealing it.

Eduardo tossed his duffel bag into the trunk and slid into the backseat. Dustin was driving, and Sean Parker was nowhere to be seen. Mark explained that Sean had gone on ahead to the party in his

BMW i series, and had already reserved them a VIP table. He'd left their names with the doorman, so they'd have no problem getting in.

Which was all good, because it gave Eduardo time to reconnect with Mark on the drive over from the airport.

Mostly, it was him talking while Mark listened—the usual nature of their relationship. He detailed the Y2M deal, and the progress he'd made with other potential advertisers. He talked a bit about some possible financing plans, about some ideas for getting more from local advertisers in each of thefacebook locations. Then he told Mark about his crazy girlfriend, and how she had left twelve new messages during the flight from New York.

Mark seemed to take it all in, but his one-word responses didn't tell Eduardo much about what he was really thinking. His update on his own progress, on what had been going on in California for the past month, on Sean Parker and the interns and the scene was his usual: "It's been interesting." Which wasn't helpful at all.

Meanwhile, the city flashed by as they made slow progress through the congested, narrow streets of the glittering city on the hill. Eduardo thought it was one of the most beautiful places he'd ever seen, but strange, too—how the houses seemed to be built right on top of one another; how the winding, curving streets—some with cobblestones and wires for cable cars—ran up hills that were almost mountains in angle and height; how you went from one corner that looked as opulent and quaint as a postcard, to another, where a gang of shambling homeless stood around a burning trash can.

And pretty soon, it was more homeless and less opulence as they passed below Geary and entered the heart of the Tenderloin district. The club was beyond O'Farrell, located in the center of a seedy stretch of check-cashing joints, fast-food restaurants, and massage parlors. As they pulled up to the nondescript entrance, Eduardo saw a huge line outside and a large man in a black suit with a headset by the door.

"This looks promising," he said as Dustin parked the car next to a pile of trash that seemed to swallow a good portion of the curb. The homeless men nearby didn't give their car a second glance. "A lot more girls in line than guys. That's a good sign."

They got out of the car and approached the front door to the club. As usual, Mark kind of hung back, so Eduardo took the initiative and walked up to the large man with the headset. The man eyed him—taking in his jacket and tie—and then glanced at Mark and Dustin, dressed like computer programmers, standing a few feet behind. The look on the man's face said it all. *These kids think they're getting in here?* It was San Francisco, sure, but even here there had to be standards. Eduardo gave him their names, and the man dutifully parroted them into his headset. Then he shrugged, surprised, and held open the door.

The place was dark and throbbing. Two floors with low ceilings, plenty of flashing strobe lights, and a Lucite stairway that curved above the bar to a raised VIP section, complete with velvet ropes and circular, leather-lined booths. The music was blaring—a mix of alternative and dance—and there were waitresses in tiny skirts and midriff-baring tops prancing through the crowd, carrying trays stacked with foofy-looking, brightly colored martinis. The place was really packed, and the waitresses were having a hell of a time keeping the martinis from toppling over.

Eduardo and his friends had made it barely ten feet into the crowd when he heard a voice over the music, from the direction of the stairs. He caught sight of Sean Parker standing midway up to the VIP section, excitedly waving at them.

"Over here!"

It took almost five minutes to work their way to the bottom of the stairs, where they had to tell another headsetted bouncer their names. Then they followed Sean up into the VIP, and joined him at one of the circular, leather-lined tables. He poured them shots from a bottle of ridiculously expensive vodka.

When they were seated and drinking, Sean launched right into a story about the last time he was in this club—with the founders of PayPal, after some awards ceremony. He talked really fast, in his usual eccentric manner, and he was so jittery—spilling his drink on the table, tapping the floor with his little, bootlike leather shoes; but Sean was always like that, Eduardo knew, his brain just ran on a faster setting than everyone else's.

While Sean talked, Eduardo couldn't help noticing the table next to theirs—because it was filled with a group of the hottest girls he'd ever seen. Four of them, to be exact, each one hotter than the next. Two blondes, in black cocktail dresses, their bare legs so long they seemed almost alien. And two brunettes, both of indeterminate ethnic origin, one bulging out of a leather bustier while the other was barely wearing a wispy summer dress that could easily have doubled for lingerie.

It took Eduardo a moment to realize that he recognized the girls—and that they were, in fact, quite *literally* the best-looking girls he'd ever seen, because they were Victoria's Secret models, right from the catalog. And then he saw something that stunned him even more: while Sean frittered on about God only knew what, one of the girls had leaned over the space connecting the two tables and was talking to Mark.

Eduardo stared in disbelief. The girl was now leaning so far forward that her ample breasts were barely contained by her bustier. Her tan skin had sparkles on it and her bare shoulders were glowing under the strobing lights. She was gorgeous. *And she was talking to Mark.*

He couldn't imagine what the conversation could possibly be about. Or how it had begun. But the girl seemed to be really enjoying herself. Mark, for his part, looked like a terrified animal caught in the headlights of an oncoming truck. But what glorious headlights they were. He barely responded, barely spoke at all—but she didn't seem to mind. She was smiling, and then she reached forward and touched Mark's leg.

Eduardo gasped. Parker was going on and on next to him. Now the entrepreneur was retelling the story of his battle with Sequoia Capital—how he believed that that crazy Welshman had forced him out of Plaxo, hired a private eye, tortured him into resigning from the company. Who knew if it was true or not, but obviously, there was really bad blood there. Sean had vowed that he was going to get back at them, someday, somehow. Then he was talking about thefacebook, how it was such an incredible thing, how he believed it was going to be the biggest thing in the world. And he seemed to really believe in it. In fact, the only thing that really bothered him about the site was the *the* in the name. It wasn't necessary. He hated unnecessary things.

On and on and on and Eduardo just sat there and listened while he kept watching Mark and the girl—

And the next thing he knew, Mark was suddenly getting up and the Victoria's Secret model had him by the hand. She led him out of the VIP area and down the Lucite stairs. And then Mark was gone.

Eduardo's head was spinning. Had he really just seen what he thought he'd seen? Could Mark really have just left the club? And wasn't he still dating that Asian girl from Harvard?

Holy shit. Eduardo was pretty sure he'd just watched Mark Zuckerberg go home with a Victoria's Secret model.

In Eduardo's mind, it was the clearest sign yet that Sean Parker was right: thefacebook was going to be the biggest thing in the world.

▸▸▸

Four days later, Eduardo was back in that window seat on the same damn American Airlines 757, his head pressed against the circular window to his right. This time there was no rain outside, but the sheets of gray were still there, vicious and violent and fierce, except this time they were in Eduardo's head, behind his eyes, grinding his thoughts like a blender on high.

Everything hurt. His body ached almost as much as his head—and he had no one to blame but himself. The past few days had been a whirlwind of business, strategizing—and drinking. Lots and lots of drinking. Beginning with that damn party, which had gone on until well past four, hours after the club had closed. Eduardo hadn't seen Mark until the next day, and Mark had been very evasive about the Victoria's Secret model. But Eduardo was certain something had happened. The harder he pressed, the more closed off Mark got—to him, a sure sign that there was something there. Eduardo could only be impressed. It felt like the world had turned upside down, and now they were deep in the rabbit hole.

Things only got crazier after that. Sean had set up a number of dinners, meetings, and cocktail outings for the time that Eduardo was there, with VCs, software reps, anyone with deep pockets who seemed interested in thefacebook. It turned out, there were a lot of people interested. In fact, they were being ferociously courted by all the major players in town. Something had certainly changed, and now there were real offers being bandied about, numbers in the many millions being whispered in their ears.

And the wining and dining was beyond excessive. They were

brought to the nicest, most expensive restaurants in San Francisco; often, the interested parties sent limos for them, or had them picked up in gleaming SUVs. When Mark couldn't get his Craigslist car to start one morning, and ended up making them late for a breakfast meeting, the VC whom they were supposed to meet had offered to buy him an SUV. Eduardo knew the man was serious—the next time he came out, he fully expected to see Mark in a new car.

But the weirdest meeting had to have been the one just the night before Eduardo's flight back to New York. He and Mark had been invited onto the yacht of one of the original founders of Sun Microsystems. It turned out, the man was an exotic eater—known for his tastes in bizarre, exotic foods. After they'd talked business for a few hours, one of the boat's staff had brought out a gleaming silver tray. On the tray was a piece of fibrous-looking meat. Eduardo had been afraid to ask—but the man had volunteered the information right away. The meat was koala—which wasn't just exotic, but, he believed, illegal. Still, it would have been rude to turn the dish away.

Sitting on the plane, waiting for the engines to come on, Eduardo still couldn't believe it all. He'd eaten koala on a yacht. He'd gotten drunk in some of the poshest places in Northern California. And he'd been whispered numbers that would make him and Mark rich, really rich.

Whatever the numbers were, though, Eduardo knew that they weren't going to sell thefacebook. In his mind, it was way too early for that. He knew that thefacebook was going to be worth a lot more in the future; hell, they were closing in on five hundred thousand members, and it was growing every day. So what if they weren't making any money? So what if, in fact, they were getting into some serious debt, barely kept alive by the eighteen thousand he'd invested into the bank account? He didn't want to sell. Mark didn't want to sell. Sean

Parker—well, who cared what Sean Parker wanted? He wasn't a member of the management team. He was an adviser. He wasn't involved. He was nobody.

Eduardo grimaced, as a new wave of gray moved through his head. Then he felt a familiar vibration, and realized that once again, he'd forgotten about his damn phone.

He yanked the thing out of his pocket. He saw that he had an incoming call—from Kelly, of course, whom he'd pretty much avoided talking to since he'd been in California.

He thought about putting the phone back in his pocket, but he knew he had a few minutes before takeoff, so he figured now was as good a time as any.

He hit the receive button and put the phone to his ear.

She was sobbing on the other end of the line, and there were loud sirens in the background. Eduardo's eyes widened, and he perked up in his seat.

"What the hell is going on?"

She spoke quickly, through her sobs. When he hadn't called her after a couple days in California, she'd done what he'd told her to do—she'd found the present he'd left for her in the closet of her dorm room. Then she'd lit the fucking thing on fire. Along with most of his clothes, which he'd left behind in her drawers. Her entire dorm room had nearly gone up. The fire department had been called, and they had sprayed the place down with fire extinguishers. Now they were even talking about arresting her.

Eduardo closed his eyes, shaking his head. Wonderful. It was just one of the joys of having a crazy girlfriend.

You never knew what she was going to do next.

CHAPTER 23 | HENLEY ON THE THAMES

Two seconds.

The difference between being a champion and being forgotten, between etching your name on a plaque and a trophy and a wall—and going home with nothing but a ribbon and some memories.

Two seconds.

Tyler felt his body sagging as he leaned forward, exhausted, his callused hands loosening against the now impotent oars. The eight-man scull was still skimming the water, still moving forward at almost racing pace—but the race was already over. Even if he hadn't seen it himself—the Dutch boat nosing them out by those bare two seconds—he would have known the results from the cheers coming from the banks of the river on either side. Those were Dutch voices shouting out to their friends and teammates, not the small contingent of Americans who had traveled halfway around the world to watch Tyler and his brother row.

Deep down, he knew that just participating in the Henley Royal Regatta was an honor, and an experience he would carry with him for the rest of his life. The event had been running annually since 1839, and took place on the longest natural straight stretch of water in England—a one mile, 550-yard section of the Thames, located in the quaint, medieval town of Henley, which dated all the way back to 1526.

The town itself was something right out of a fairy tale. Some of the original buildings still stood, and Tyler and his brother had spent much of the five-day event wandering the narrow streets with their host families, hitting the pubs, churches, shops—well, mostly the pubs.

But despite the culture they'd experienced during the week, they'd come to Henley for one reason: to race in the Grand Challenge Cup, against the best crew in the world. And despite their best efforts, they'd come up short.

Two lousy seconds short.

▸ ▸ ▸

By the time they'd climbed out of the scull and onto the dock for the award ceremony, much of the high-profile audience had streamed out of the Stewards' Enclosure—a sprawling, overly prestigious viewing area that you had to be a member or a member's guest to enter—and were milling about, waiting for Prince Albert to do the honors. The prince seemed much shorter in person, but Tyler was quite impressed when the royal shook his hand and seemed to know his name from memory. The mere fact that Albert was there was a bit of good luck; usually, it was a lesser royal doing the award duties, but Albert had made the trip from Monaco in honor of his grandfather, who had been one of the premier rowers of his day—although Jack Kelly had, ironically, been banned from competing in Henley because of his bricklayer background, which Albert now made up for by hosting the event itself.

But a handshake was all Tyler and Cameron received from the dashing prince; the real trophy went to the Dutch team, who took the honor graciously. It was a bit bitter, watching the other crew hefting the trophy above their heads, but Tyler was a good sport, and he applauded along with the rest of the crowd.

Afterward, he and Cameron wandered into the Stewards' Enclosure—they had been given badges by their host family, who were members—and spent the next few minutes admiring the sometimes bizarre fashions of the British rowing fans; the brightly colored jackets and ties, the long, flowing dresses, the summer hats—the works. It was the first week of July, and the sun was beaming down, but nobody seemed to notice the heat. Maybe that was because there were four bars in the Enclosure, as well as a covered luncheon area and tea tent.

"Can't win 'em all. Nice job, boys. Down by just a nose."

Tyler forced a smile as he spotted their host father near the back of the Enclosure, who was separating himself from a group of his friends and hobbling toward them. The man was pudgy, midfifties, and had bright red cheeks set off from a pug nose and deep-set blue eyes. The amiable man made his living as a barrister in London—just a thirty-five-mile commute away—but had been a rower himself for Oxford twenty-five years earlier. He hadn't missed a Henley since, and had been hosting crew members from across the pond for nearly a decade.

"Thanks," Tyler responded, trying to sound upbeat. "It was a tough one. But they deserved it. They worked harder."

And Tyler was pretty sure he meant it as he said it. Crew races weren't usually that close, and for the Dutch team to pull it out by two seconds—as clichéd as it sounded, it was simply a matter of who had wanted it more.

"Well, my daughter took some wonderful pictures," the barrister said. "But she's gone home now, unfortunately."

"Maybe she can e-mail them to us," Cameron chimed in. Someone they didn't know handed each of them a smoked-glass mug filled with warm beer. It was a tough tradition to get used to—but Tyler and Cameron had been working at it since they'd arrived in Henley.

"Well, are you boys on thefacebook?"

Tyler froze, the mug of beer pressed against his lips. He wasn't certain he'd heard the man right. Sure, he'd heard a lot of people talking about that damn Web site over the past couple of months—but never in an English accent. He would never have expected to hear it mentioned in a medieval British town on the banks of the Thames.

"Sorry?" he stammered, hoping he really had just misheard.

"You know, the Web site. My daughter tells me all the college kids in America are using it. She's just returned from a year abroad, you know, at Amherst. And she's on that Web site all the time. I'm sure you can find her there, whenever you want, and she'll e-mail you the pictures."

Tyler glanced at his brother. He could see his own feelings reflected in Cameron's eyes. Even here, across the ocean, thousands of miles from Harvard—they were talking about thefacebook. Even though it was still only available to college kids in the United States—and how many colleges? Thirty? Forty? Fifty? It was exploding in ways none of them could have foreseen.

And meanwhile, ConnectU had pretty much stalled at the gate. Despite the fact that ConnectU was chock-full of features, had launched in a number of schools at the same time—it simply couldn't compete with the viral nature of thefacebook. Whether it was the curse of first-mover advantage, or simply that people liked thefacebook better, ConnectU was nothing but a little blip on the social networking radar.

Thefacebook was a relative monster. Godzilla, crushing everything in its path.

Tyler forced a smile back on his lips, and made some small talk with the barrister, pushing the subject of thefacebook aside—but all the while, his mind was churning through thoughts that he'd been fighting for the past four weeks.

He, Cameron, and Divya had tried to get beyond the anger and frustration—had tried to make the best of a bad situation. And it had gotten them nowhere. They'd launched their site, they'd gone after thefacebook's audience in a number of ways—and they simply couldn't compete. College kids were going to join the social network that their friends were already on, not something new they'd never heard of. Thefacebook was stomping all competitors into the ground.

The truth was, they'd been beat. Harvard had washed its hands of the situation. Mark had ignored their e-mails and their cease-and-desist letter. There was really only one option left. Larry Summers had practically spelled it out for them—and yet, so far, it was something they had resisted.

Tyler and Cameron knew a bit about lawsuits from their father's business; Wall Street was brimming with lawyers, and they had heard many war stories from the world of the corporate courts. They knew that a lawsuit was an ugly thing, no matter how it eventually panned out. It was an act of last resort—but wasn't that exactly where they were? The last resort? Beaten by two seconds by a kid with a computer—a kid who showed no remorse, who had left them no choice.

Tyler also knew that it wasn't just the legal process that was going to get ugly; he could imagine how things were going to play out in the press. He had always been pretty self-aware—and he could guess what people were going to say, picturing him and his brother next to Mark Zuckerberg. Hell, the *Crimson* had already attacked them in a number of editorials; in fact, one writer had even called them "Neanderthals." The writer of that piece, it had turned out, had been a girl who had once dated one of Tyler's Porc brothers and had spent their entire relationship nagging the poor kid about the "evil" nature of the Final Clubs. But she was indicative of what they would face if they launched a lawsuit against Mark Zuckerberg.

If this were an eighties movie, Tyler and Cameron would certainly be the bad guys. They'd be dressed as skeletons, chasing the Karate Kid around a school dance. They were jocks from a wealthy, tony family. Mark was a nebbishy geek who had hacked his way to stardom. This was a class battle the journalists couldn't ignore: rich, privileged kids who believed the establishment existed to protect their rights, against a hacker who had been willing to break the rules. Honor code vs. hackers code.

Tyler knew how he and his brother were going to look.

But if that's what it would take to have even a fighting chance at finding justice—they were willing to put on the skeleton costumes and give it a go.

Mark Zuckerberg hadn't left them any choice.

CHAPTER 24 | JULY 28, 2004

Eyes closed.

Heart pounding.

Sweat streaming down the skin of his back.

Eduardo was angry, that we know for certain. Where he was—wandering the streets of New York in a bitter haze, or trapped on a subway, hurtling forward at thirty miles per hour, his arms wrapped tightly around a sticky chrome pole, his body jerking forward and back as the crowd of strangers pressed into him from every side, we can't know for sure. But wherever he was, he was fuming—and he was about to do something that would change the course of his life.

It had all started about three days before. At the time, Eduardo had actually been on an emotional high; since he'd gotten back from California—and quickly broken up with Kelly, nipping her unbalanced theatrics in the bud—things had been going really well in New York, and he was feeling good about the progress he had been making with Y2M and the other advertisers he'd lined up for the Web site. So he'd dialed up Mark in the La Jennifer Way house to report to him—and that's when things had started to go downhill.

To say that Mark had been unappreciative of Eduardo's hard work in New York would be an understatement; in Eduardo's view, Mark barely listened at all as Eduardo explained what he'd gotten done, and immediately launched into some story about a party Sean Parker had brought them to the night before, something involving a Stanford sorority and a truckload of Jägermeister.

After that, the conversation had devolved into Mark's usual refrain of late—that Eduardo should move out to California, because that's where it was all happening. The computer coding, the networking with potential investors, the meetings with VCs and software honchos—Mark pretty much intimated that Eduardo was wasting his time in New York, when everything that thefacebook needed could be found right there, in Silicon Valley.

Eduardo had tried to point out that New York was also an important center for the things a growing start-up needed—from advertising dollars to banking contacts—but Mark hadn't really wanted to listen to him at all. And then, to make matters worse, Sean Parker had jumped on the phone, and had immediately started talking about two potential investors whom he was going to introduce to Mark. In fact, Parker had said, these investors were ready to put up real money—and if Mark liked them, and they liked Mark, it would happen pretty fast.

Eduardo had nearly lost it, right there on the phone. He'd quickly explained to Parker that he was running the business side of thefacebook, that any meetings with investors would have to include him—and why the hell was Parker setting up these sort of meetings anyway? In Eduardo's mind, it wasn't even Mark's job to be looking for potential investors; he was supposed to just run the computer side of the company. And Parker wasn't involved at all. He was a houseguest. That's it. A fucking houseguest.

After that first phone call, Eduardo's emotions had started to shift from frustration to pure anger. So he'd done something impetuous—maybe out of that anger, or maybe because at the time it had seemed the proper thing to do. To clarify his feelings, and let Mark know that it wasn't kosher to leave him out of the loop.

He'd crashed out a letter reiterating his and Mark's business relationship; specifically, he'd respelled out the agreement they'd made when they'd started thefacebook, that Eduardo was in charge of the business side of the company, and that Mark was supposed to be out in California working on the computer code. Furthermore, Eduardo had added that since he owned 30 percent of the company, he had the power to keep them from accepting any financial deals that he did not agree with. Mark had to accept that reality—and Eduardo wanted written confirmation that he could run the business side of things as he saw fit.

Eduardo had known when he'd written the thing that it wasn't the sort of letter that a guy like Mark Zuckerberg would react well to—but Eduardo had wanted to be as clear as possible. Sure, Sean Parker had taken them to some cool parties, maybe even helped get Mark laid with a Victoria's Secret model—but in Eduardo's view, he wasn't involved in thefacebook. Eduardo was the CFO, he'd put up the money that had made thefacebook possible, he was still the one funding their adventure in California—and even though he was in New York, he was still supposed to be calling the shots.

After receiving the letter, Mark had left him a bunch of messages on his voice mail—more entreaties for Eduardo to move out there to

California, more stories about how great it was out there, more reassurances that everything was going great with the company and there was no reason for them to bicker about stupid things that didn't matter anyway—in his bizarre worldview. Finally, Eduardo had called him back, just a little while ago—and things had gone from bad to worse.

Mark had told him that he'd met the two investors Sean Parker had told Eduardo about, and they were really interested in making an angel investment—basically giving thefacebook some money so it could continue growing at the same rapid rate. Thefacebook needed the money, since it was beginning to fall into serious debt; the more people who were signing up, the more servers that were needed to handle the traffic—and soon they were going to have to hire more people to handle everything that was going on.

But to Eduardo, that was all beside the point. In his opinion, Mark had deliberately ignored the sentiment of his letter—and was taking business meetings without Eduardo being present. He wasn't simply stepping on Eduardo's toes; he and Sean Parker seemed like they were trying to cut off Eduardo's feet.

Maybe Mark didn't think Eduardo was serious, that the letter had been just a method of letting off steam. And maybe it was, in a way. But Mark's attitude was really pissing Eduardo off; in Eduardo's opinion, they were out there, living it up in California on Eduardo's dime. The house in California? The computer equipment? The servers? It was all coming out of the bank account that Eduardo had opened, as far as Eduardo was concerned. That Eduardo had financed from his own, personal funds. Eduardo was paying for everything, in his mind, and Mark was ignoring him. Treating him like an angry girlfriend that he just didn't give a shit about anymore.

Maybe Eduardo was overreacting—but now, three days later,

fuming somewhere in New York—he was growing more and more certain that he had to do something to show Mark exactly how he felt.

He had to send a message—one that Mark couldn't simply ignore.

▶ ▶ ▶

We can picture what must have happened next: Eduardo spinning through the revolving glass door of a midtown Bank of America office, his face a mask of pure determination, his oxford shirt soaked with sweat from either a subway ride or twenty minutes trapped in a traffic-bound cab.

He moves right past the teller stations that run along one side of the wide, rectangular front area of the bank and heads directly to one of the branch associate cubicles. By the time the balding, middle-aged banker gestures him into a seat and asks what he could do for him, Eduardo has already pulled his bankbook out of his pocket. He slams the little booklet onto the desk in front of the man and gives him his most serious, adult stare.

"I want to freeze my bank account. And cancel all existing checks and lines of credit attached to this account."

As the man begins the process, assuredly Eduardo feels a burst of adrenaline move through him. He must know he is crossing a line—but this was going to send Mark a real message, let him know how serious Eduardo is. Really, in Eduardo's mind, it is Mark's own fault that Eduardo even has the power to do such a thing—when Eduardo had first opened the Bank of America account for thefacebook, he'd sent Mark the necessary forms to become a cosignatory on the account, along with the blank checks that were funding his California lifestyle. Mark, being Mark, had never filled out the paperwork. Nor had he ever put any of his own money into the company. He'd been perfectly content to live off of Eduardo's funds. As if Eduardo were his own, personal banker. His partner—except, now, he had started to make decisions without Eduardo's involvement, and Eduardo had to let him know that it simply wasn't okay. Eduardo had to let Mark know what it meant to be a good partner. Eduardo didn't care if every thefacebook page was a Mark Zuckerberg production. But the company itself was the result of a combined effort. Eduardo was a businessman, and this move is all business.

As Eduardo watches the banker hit the necessary keys on his computer to freeze thefacebook's bank account, maybe he wonders, for the briefest of seconds, if he is going too far. If he does, he can cancel the thought with another: a picture of Mark and Sean running around California in Parker's BMW, taking meetings with investors, maybe even laughing at Eduardo's efforts to rein them in.

They wouldn't be laughing when they tried to cash the next blank check—that was for sure.

CHAPTER 25 | SAN FRANCISCO

This time, the revolution wasn't going to begin with a bang.

Instead, Sean Parker realized, it was going to start with the whir of a state-of-the-art elevator, speeding up the spine of a massive, San Francisco skyscraper—and the sickly, soft chords of a brutally mangled Beatles song, pumped through speakers embedded above the fluorescent lights that lit the carpeted, cubic lift.

Sean had to admit, there was something strangely poetic about the setting; this was potentially the beginning of the next great digital-social seismic change, and the only thing that marked the seconds ticking away toward that epochal event was the horrific beat of canned Muzak.

He stifled the urge to grin as he stood next to Mark in the center of the otherwise empty elevator, staring up at the little glowing numbers that tracked their progress up the skyscraper. At the moment, they were somewhere between the ninth and tenth floors of the fifty-two-story building, moving upward at an incredible pace. Sean felt his ears pop from the change in altitude—which was a good thing; for the briefest of seconds, he couldn't hear the Muzak, which allowed him to order his thoughts—or at least corral them in as close to a semblance of order as his highly energized gray matter would allow.

Things were happening quickly—much faster, even, than Sean himself had expected. He'd only just a few weeks ago moved in with the eccentric genius standing next to him in the elevator—and now here they were, on their way to a meeting that could very well launch them into a partnership that would change the face of the Internet itself—and put them well on their way toward the billion-dollar payoff Sean had envisioned when he first saw thefacebook in that dorm room on the Stanford campus.

Sean glanced toward the twenty-year-old kid standing next to him. If Mark was nervous, he didn't show it. Or more accurately, he didn't look any more uncomfortable or anxious than usual; his face was a mask of indifference, his eyes trained on those same ascending numbers above the elevator doors.

Since they'd run into each other on the street outside of Palo Alto, Sean had gotten to know the eccentric kid pretty well, and he was genuinely beginning to like him. Certainly, Mark was strange; socially awkward didn't begin to describe his standoffish mannerisms. But even despite the walls the kid had built around himself, Sean could tell that his initial opinion of the boy genius was not far off. Mark was brilliant, ambitious, and had a caustic sense of humor. For the most part, he was a quiet person; Sean had taken him to numerous parties, but Mark was never comfortable at any of them—he was much happier lodged in front of his computer, sometimes twenty hours at a stretch. He still had that college girlfriend whom he saw about once a week, and he liked to take long drives when he got tired of the computer—but otherwise he was a coding machine. He lived, breathed, and ate the company he had created.

Sean could not have asked for more from a fledgling entrepreneur; in fact, sometimes he had to remind himself that the kid standing next to him was barely twenty years old. His lifestyle was still somewhat immature, but his focus was amazing, and Sean was certain he was willing to make any sacrifice necessary to continue growing his Web site; which was exactly why Sean felt certain that the step they were about to take was the right one. That the meeting they were hurtling toward would be the catalyst to that billion-dollar payoff that had eluded him through two successful start-ups and half a decade navigating the busts and booms of a newly reemergent Silicon Valley.

In a weird way, Sean had Eduardo Saverin to thank for pushing things to a head so rapidly; if it hadn't been for Eduardo's actions over the past couple of weeks, it might have taken an entire summer to get Mark to this point. But Eduardo had done the job of pushing Mark to make a big move forward for Sean—in the most bizarre, and unexpected fashion.

First, there had been that idiotic letter. Sean thought it was like a kidnapper's ransom letter, really—it might as well have been written in cut-out words from newspapers and colored magazines. Threatening, cajoling, demanding—the kid had some serious self-awareness issues he needed to deal with. The very idea that he was running the business side of an Internet company from New York while the rest of his partners were actually building the site out in California was the height of absurdity. And then, trying to hold his 30 percent ownership over Mark like it was some sort of weapon—Eduardo had gone right off his rocker.

Still, Mark had tried to be reasonable with his friend—and Sean had been right there with him, trying to smooth things over. There was really no need to turn the letter into more than it was—a desperate, childish plea to be more included in what was going on with the company, which Mark certainly could have accepted.

But before Mark and his friend had worked anything out, Eduardo had gone and crossed the line: he'd frozen the company's bank account, effectively cutting Mark and Dustin off at the throat. With that single act, he'd taken a shot at the soul of the company itself. Whether he realized it or not, his actions could easily have destroyed everything Mark had worked on—because without money, the company couldn't function. If the servers went down for even a day, it would hurt the reputation of thefacebook—possibly in an irrevocable way. Users were fickle; Friendster had proven that fact time and again. If people decided to leave the Web site—well, then that could quickly have turned disastrous. Even a small-scale exodus would reverberate through the whole user base, because all of the users were interconnected. College kids were online because their friends were online; one domino goes, a dozen more follow.

Maybe Eduardo hadn't really understood what he was doing; maybe he'd acted out of anger, frustration, God knew what—but simply put, in Sean's view, his childish maneuver had made it difficult for him to stay a big part of the company going forward. And in Sean's mind, it really had been the act of a child, not the businessman Eduardo saw himself as. Like a little kid on the playground, screaming at his friends: "If you don't do things my way, I'm taking my toys and going home!" Well, Eduardo had taken his toys—and now Mark had made a decision that was going to change thefacebook in ways Eduardo couldn't imagine.

First, under Sean's guidance, Mark had reincorporated the company as a Delaware LLC—to protect it from Saverin's whims, and also to begin the restructuring that Sean knew would be necessary to raise the money the company needed to go forward. At the same time,

Mark had gathered what resources he could, and put his own money into keeping the company alive for the moment until they could set things right. Drawing from his own college savings, money that had been earmarked for his tuition, Mark had managed to come up with enough to keep the servers running for the time being; but the company was rapidly heading toward real financial trouble, something Mark could no longer ignore.

Furthermore, it wasn't just the servers or the need for new employees that was going to be a problem. To add to everything else, just a few days before, they'd received a letter from a law firm that had been hired by the ConnectU founders—the Winklevoss twins, the jocky seniors who had hired Mark to work on some dating site back when he was still in school. The letter was the first step in the initiation of a lawsuit—a sort of warning shot at thefacebook's bow, as Sean saw it.

Even before the letter from the law firm, Sean had spoken at length with Mark about the ConnectU situation, and he'd also done some research of his own into the situation. In his mind, the Winklevoss twins were a nuisance, but not a real danger to the future of the company. A mild concern, at best; in Sean's opinion, their claims were unfounded and overblown. So Mark had done a little work for their dating site before coming up with the idea for thefacebook? So what? There were a hundred social networks out there; every computer geek in every dorm room was working on some program like thefacebook; that didn't mean they were all subject to lawsuits. And all these social networks were pretty similar at their core. Mark's own argument—that there are an infinite number of designs for a chair, but that doesn't mean everyone who makes a chair is stealing from someone else—seemed as good as any to Sean. If anything, they were all borrowing from Friendster when it came right down to it; the

ConnectU twins hadn't exactly invented the wheel, that was for sure. Mark had done nothing wrong, nothing that every other entrepreneur in the Valley hadn't done a dozen times before.

Even so, if the twins persisted—and the legal letter seemed to indicate that they would—it was going to cost Mark upward of two hundred thousand dollars to defend himself. Which meant he needed to raise more money—fast. And since selling the company wasn't an option—not in Sean or Mark's mind, that was certain—they needed an angel investment to tide them over until they could reach a valuation that would make all these problems seem petty and insignificant. Sean only wished he had that sort of money—but the way things had worked out with Napster and Plaxo, Parker didn't have anywhere close to what Mark would need to keep thefacebook afloat.

So instead, Sean did what he did best; he made a connection—one that he was pretty sure was going to be the key to what had to happen next, to make thefacebook into what he knew it could be.

Watching the numbers sprinting upward as the elevator brought them closer and closer to their goal—Sean knew that once again he'd done exactly the right thing. All Mark had to do was ace the meeting—and they were on their way.

He threw another sideways glance at the boy wonder—and again, got nothing in return. He reminded himself that Mark's silence didn't mean anything. The kid would be able to perform when the time came. All Sean needed from him was fifteen minutes.

"You know they filmed *Towering Inferno* here, right?" Sean said, trying to keep the mood in the elevator light and easy. He thought he saw the slightest sliver of a smile on Mark's lips.

"That's comforting," Mark robotically responded. Sean was pretty sure he was being ironic, and he allowed himself that grin he'd been fighting.

It really was a fitting place for the meeting—not because of the movie, but because it was one of the most impressive landmarks in the city. Formerly the Bank of America Center, the behemoth at 555 California Street was an architectural wonder, an enormous, polished granite tower with thousands of bay windows that could be seen for miles, a 750-foot spire rising right out of the epicenter of the city's financial district.

And the man they were on their way to meet—well, he was nearly as impressive as the building itself, both in personal reputation and in what he had already achieved.

"Peter's going to love you," Sean responded. "Fifteen minutes, in and out, that's all it's going to take."

Deep down, he was certain that he was right. Peter Thiel—the founding force behind the incredibly successful company PayPal, head of the multibillion-dollar venture fund Clarium Capital, former chess master, and one of the richest men in the country—was intimidating, fast-talking, and a true genius—but he was also exactly the sort of angel investor who had the guts and the foresight to understand how important—how groundbreaking—thefacebook had the potential to be. Because Thiel, like Sean Parker and Mark Zuckerberg, was more than just an entrepreneur: he saw himself as a revolutionary.

A former lawyer from Stanford, Thiel was a well-known libertarian; during law school, he'd founded the *Stanford Review*, and he was a firm believer in the value of the free exchange of information that thefacebook celebrated within its social networks. Though secretive and incredibly competitive, Thiel was always searching for the next big thing—and Sean knew that he shared his own interest in the social networking space.

Sean had never worked directly with Thiel before, but he'd been involved in getting Thiel invested minorly in Friendster, and he'd always kept the former PayPal CEO in the back of his mind, in case another opportunity ever arose.

The opportunity had arisen—and was still rising, floor by floor, toward Thiel's glass-and-chrome office, where Thiel—along with Reid Hoffman, his colleague from PayPal and also the CEO and cofounder of LinkedIn, as well as Matt Kohler, a brilliant engineer and rising Valley star—were waiting to hear the pitch from the quirky kid who'd lately been taking the Internet world by storm.

If Thiel liked what he heard—well, Sean could think of no better way to put it: the revolution that was thefacebook would truly, and earnestly, begin.

▸ ▸ ▸

Five hundred thousand dollars.

Three hours later, the number reverberated through Sean's skull as he stood in near silence in the rapidly descending elevator next to Mark, watching those same glowing numbers count back down as they hurtled back toward the lobby of the great, granite building at 555 California.

Five hundred thousand dollars.

In the general scheme of things, of course, it wasn't a huge number. It wasn't life-changing money, it wasn't empire-making money, it wasn't fuck-you money—it wasn't even as much money as Mark had once turned down, back in high school, when he'd created that MP3 player add-on, simply because he didn't really give a damn about money, whether it was a thousand dollars borrowed from a friend to start a company, or a million dollars thrown his way from an even bigger company. As far as Sean could tell, Mark still didn't really give a damn about money; but he couldn't ignore the sentiment that came with those five hundred thousand dollars, the promise of a future for the company that he'd started in that Harvard dorm room.

Peter Thiel had been exactly everything that Sean had prepared Mark for. Scary as hell, brilliant as hell—and willing to play ball. More than that, he'd turned a fifteen-minute pitch meeting into a lunch and an afternoon spent going over the details—of the deal that would ensure thefacebook's survival, once and for all. At one point, Sean and Mark had even been sent out of the meeting, to walk around town while Thiel and Hoffman and Kohler discussed their pitch—but by the end of the afternoon, Thiel had given them the great news: thefacebook was on its way.

Or, as the company was now going to be called—just "Facebook." Sean's idea, because he'd been so damn annoyed by that *the* in the Web site's name, he'd finally gone and gotten Mark to slice it right off in the reorganization that was now an inevitability, a necessary step in getting that five-hundred-thousand-dollar "angel" investment that was going to save all their necks.

Seed money, Thiel had called it. Enough to get them through the next few months—and along with it, a promise of more when the time came, when the need arose. In exchange, Thiel was going to get about 7 percent of the newly formed company, and a seat on the five-man board of directors that would lead the company going forward. Mark would still control the majority of the seats, and thus the company itself. He'd also keep the lion's share of the company's stock, even in its new iteration. But Thiel would become a guiding force, leading them forward along with Sean and Mark. It simply didn't get any better than that.

Standing there in that elevator, listening to the Muzak—some Rolling Stones bastardization that made Sean want to vomit on the inside—it was an overwhelming moment. Still, Sean knew that there was work to be done; he knew that this re-formation of the company was going to create a pretty intense situation.

Reincorporating was necessary, both Thiel and he had agreed. Facebook had to become a new entity, shedding its dorm-room genesis and moving into a sort of "New Testament" status. They were going to have to reissue shares to represent the new setup, to include Thiel and of course Sean himself—who'd been working as a partner to Mark since he'd moved into the house anyway—and Dustin and Chris.

Which left the question of Eduardo. Initially, Mark had decided, and Sean had agreed, Eduardo would still get his 30 percent. The intention was to include Eduardo and involve him as much as he wanted to be involved. But the new corporation would have different rules—it *had* to have different rules. There just wasn't any way to run a business without the ability to issue more shares as was necessitated by the evolving situation. Going forward, people had to be given shares based on the amount of work any particular individual gave to the company. This wasn't some dorm-room project anymore, this was a real company, with a real investor. People had to be reimbursed as if this was any other company, because otherwise it would be impossible to create a real valuation based on what Facebook achieved.

Which meant that if Mark, Dustin, and Sean were doing all the work to make the company successful, they would get issued more shares. If Eduardo was in New York, working on finding more advertising partners—he would get shares accordingly. But if he didn't produce, well, he would be diluted, just like anyone and everyone else. Hell, if they needed to raise more money in the future, they would all be diluted.

From Sean's point of view, Eduardo had done a horrible thing; he'd threatened the very company during its most fragile stage. Mark didn't seem to hate Eduardo for it—Mark didn't have the capacity, or the interest, to hate anyone. But in Sean's view, Eduardo had shown where he stood. To Mark and Dustin and Sean, Facebook was everything. It was their lives.

In fact, Mark had told Thiel in the meeting that he'd probably not even return to Harvard when the summer ended; he was going to stay in California and continue the adventure. He'd take it month by month—but if Facebook kept progressing, he didn't envision returning to Harvard anytime soon. Like Bill Gates had said: "If Microsoft didn't work out, he could always go back to Harvard."

Sure, if Facebook didn't work out, Mark could always go back to school—but Sean doubted he ever would. He was going to continue his endless summer; and most likely, Dustin would stay on in California as well.

But Eduardo? Well, from what Sean knew of the kid, Eduardo would never quit school. He'd already proven that he wasn't going to give up everything else for Facebook. That simply wasn't who he was. He had other interests. For instance, back at Harvard, from what Sean understood, he had the Phoenix. In New York, he'd had that internship, even though he'd quit in the first week.

Eduardo would go back to school. But Mark Zuckerberg had found his place in the world.

Sean watched the numbers descending, the excitement finally starting to die down inside him. He forced his pulse to return to a steady beat, like the steady bytes and bits of a processing computer hard drive.

He knew that there were still obstacles ahead. So much work still to be done.

First and foremost, Mark would probably have to get Eduardo to agree to the legal details—just to make things cleaner, from a lawyer's point of view. As harsh as it sounded, from a practical point of view, Eduardo should understand. This wasn't a personal issue, it was business. And Eduardo saw himself first and foremost as a businessman.

Sean and Peter were successful entrepreneurs, and they had explained to Mark how all this worked. Start-up companies like Facebook really had two distinct starting points. There was the first starting point: some kids in a dorm room, hacking around on a computer. Then there was the second starting point: here, in a skyscraper in downtown San Francisco.

If you were there in the dorm room, you had an exciting and wonderful story to tell. You got to be part of something really cool, that spark of genius, that flame bursting up out of nowhere, that lightning bolt of imagination.

If you were there in the skyscraper—well, that was something very different. That was the real beginning of the Company with a capital *C.* That was the real business, the corporation—the second lightning bolt, that really took you straight up into the heavens.

Really, it was something Eduardo should understand. It wasn't about two kids in a dorm room anymore.

And if he didn't get it? If he didn't understand? If he didn't want to understand?

Well, in many ways, if Eduardo didn't get it—then in Sean's opinion, he didn't really care about Facebook the same way they did. Then he was no better than the Winklevoss twins, trying to grasp onto Mark's ankles as he headed toward the heavens.

Either way, Mark had to know he was making the right decision for the company. Sean and Thiel had made it clear; no investor was going to hand them money with some kid running around New York, claiming to be the head of the business side of the company, flaunting some "30 percent" ownership status, holding it over them like a saber, ready to chop off their heads.

Freezing their bank account.

Threatening them.

Threatening Facebook.

That's what it all came down to—Facebook. The company. The revolution. Sean could tell, that's all Mark cared about now. He knew he was on top of something huge. This Mark Zuckerberg production was going to change the world. Like Napster, but bigger—Facebook was all about freedom of information. A truly digital social network. Putting the real world onto the Internet.

Eduardo would have to understand. And if he didn't?

Then, in the larger scheme of things, he didn't matter. He didn't exist.

Standing there in the elevator, Sean thought about the last thing Peter Thiel had said to Mark after making the deal that would take the company to the next level. Right after telling Mark that when they got to three million Facebook members, he could take Thiel's 360 Ferrari Spyder out for a drive. Right after filling out the paperwork that would enable Mark to draw on that five hundred thousand dollars in seed money—to build Facebook however he wanted, as big as he could dream.

Thiel had leaned forward over his desk, and looked Mark right in the eyes.

"Just don't fuck it up."

Sean grinned as he stared at the glowing numbers above the elevator doors.

Thiel had nothing to worry about. Sean knew his new friend. Mark Zuckerberg wasn't going to let anyone fuck Facebook up. He was going to lead this revolution—no matter what the cost.

CHAPTER 26 | OCTOBER 2004

If Eduardo had squinted real hard, maybe spun himself around a little, he could have been right back in Mark's messy dorm room in Kirkland House, watching his friend plug away at his laptop. Even the furniture in the open, central office of the new rental "casa Facebook" in Los Altos, California, looked like it had been shipped in from Harvard—scuffed wooden chairs, futons, worn desks and couches that seemed a dorm-room-chic mix of IKEA and the Salvation Army. Out back, the porch was speckled with paintball shots, and there were cardboard boxes everywhere, making it seem like they were a team of squatters more than a start-up operation in full swing. Of course, there were computers everywhere—on the desks, on the floor, on the counters next to boxes of cereal and bags of potato chips—but even with all the hardware, the house had the feel of a college dorm—which was exactly what Mark and the rest had been going for. Even though they were now working round the clock—at that very minute, Mark and Dustin were behind computer screens, plugging away, while two young men in suits—lawyers, Eduardo knew, from the firm the company had contracted to handle the new incorporation contracts, among other things—shuffled around by the door that led to the kitchen—they did not want to lose the college feel of the company, because it would always be, at its heart, a college experiment gone viral.

And despite the somewhat choreographed chaos, this five-bedroom house was still more suited for Mark and the gang than the previous one in suburban Palo Alto—not that the move had entirely been Mark's choice. After a series of complaint letters and visits from the landlord, they'd been pretty much kicked out of the La Jennifer Way sublet for, among other things, climbing on the roof, playing music too loud, throwing patio furniture into the pool, and damaging the chimney with the zip line. Eduardo had a pretty good feeling that they wouldn't be getting the security deposit back anytime soon.

Which was okay, now, because Facebook had its own financing in place; an angel investment from Peter Thiel, which was paying for this new house, all this computer equipment, more servers than Eduardo had imagined they'd ever need—and the lawyers, who had greeted Eduardo with smiles and handshakes when he'd entered the house after the long flight and taxi ride that had brought him in from Cambridge that very morning.

Eduardo had slept most of the trip; eight weeks into the new school year—his senior year—and he was already exhausted. Even though he was taking a bit of a reduced class load so that he could continue his work on Facebook, there was always so much to do at Harvard—from the thesis he was already working on for his major, to the Investment Association, which he was still a part of, and of course the Phoenix, which kept his weekends filled—especially since he was single after having broken up with Kelly. And now that it was the beginning of another punch season, it was his turn to help pick the new crop of campus social kings.

On top of all that, of course, there was Facebook.

Eduardo leaned back in his chair, which was positioned to the side of a round table that took up most of the center of the main office in the house, and watched Mark as he worked away at his laptop computer. The glow of the screen splashed across Mark's pallid cheeks, tiny sequences of code reflecting across the bluish globes of his eyes. Mark had barely greeted him when he'd first come into the house—really, just a nod and a word or two—but that wasn't unusual, nor did Eduardo read anything into it. Actually, things had been going quite well between them over the past eight weeks, since he'd been back at school.

The rocky few weeks of summer seemed almost forgotten, now; Mark had been pretty pissed off about the bank account situation, and he'd gone right ahead with the investor meetings that had led to the financing from Thiel, despite Eduardo's wishes. They'd had it out on the phone a number of times—arguing like any two friends might, who were involved in something that had gotten bigger than either of them had really expected—but they'd come to a sort of détente, finally agreeing that the important thing was the company, that it continued moving forward in a smooth fashion. Eduardo had probably overreacted with the bank account, and Mark had been a bit distant and selfish by keeping Eduardo out of the loop—but Eduardo was willing to be reasonable and move forward, for the good of the company. This was business, and they were friends; they would find a way to work things out.

To that end, Mark had asked that Eduardo step back a little—to ease his own concerns, and also so that Eduardo could focus on finishing up school. He'd convinced Eduardo that the company was getting too big for one person to try to control all the business side of things, that it was simply impossible what he'd been demanding. As things continued to grow—they were closing in on 750,000 users now, heading toward a million!—Mark and Dustin were taking time off from college, maybe a semester, probably not longer—and they were also planning on hiring a sales executive to pick up the slack, handle some of the things Eduardo had been working on in New York. They were also rapidly adding functions to the site—some of them quite incredible. They'd created something called a "wall," where people could communicate with one another in a very open format that hadn't really been seen before on any social network. And there were now groups available for people to join and create—an idea Eduardo had talked about with Mark back when they were first coming up with the site. The pace of invention was just incredible, almost mimicking the viral growth of the user base.

In the end, Eduardo, after having calmed down a bit from his burst of anger back in July, had come to the conclusion that Mark was going to do things Mark's way; and now that the summer was over and Eduardo was back in school, he was probably better off anyway. The important thing was that the company was thriving. With Thiel's money, Eduardo wasn't risking his own cash anymore; and really, Thiel was a bottomless pit, so there was no risk that the company wouldn't be able to handle whatever was thrown its way.

For Eduardo's part, he was actually glad to be back at school. One of the great thrills of his senior year had been the first week; he'd heard, through friends at the Phoenix, that President Summers had announced to the entering freshmen that he had checked them all out on Facebook. It was a pretty incredible thought—that the president of Harvard was using their site to get to know the incoming class. Just ten months earlier, Mark and Eduardo had been two geeky nobodies, and now the president of Harvard was name-checking their creation.

In light of that, did any of the squabbling between him and Mark really matter? When Mark had called and asked him to come out to California to sign some papers—basically, some new incorporation stuff, for the new restructure of the company now that Thiel was on board—Eduardo had shrugged, figuring it was all for the best.

So, as one of the lawyers wandered across the central office and handed him a stack of legal papers, he took a deep breath, glanced at Mark again—then started reading through the legalese.

From a first glance, it was pretty complicated stuff. Four documents in all, numbering many pages altogether. First, there were two common-stock purchase agreements—essentially, allowing him to "buy" stock in the newly reincorporated "Facebook," instead of the now worthless "stock" he had in the old thefacebook. Second, there was an exchange agreement, for exchanging his old shares of thefacebook for new shares in the new company. And last, there was a holder voting agreement, something Eduardo didn't entirely understand, but seemed like more legalese that was necessary for the new company to function.

The lawyers did their best to explain the documents as Eduardo leafed through them. After the repurchases and the exchange, Eduardo would have a total of 1,328,334 shares of the new company. According to the lawyers—and Mark, who looked up a few times from his computer to help outline the new structure—Eduardo would thus have about 34.4 percent ownership of Facebook at the moment—the rise in his share percentage from the original 30 percent due to the necessity, in the future, of dilution as they hired more people and awarded other investors that would surely come along.

Mark's own percentage had gone down to about 51 percent, and Dustin now owned 6.81 percent of the company. Sean Parker had been given 6.47 percent—more than he deserved, in Eduardo's mind—and Thiel had what worked out to around 7 percent.

Included in the documents was a vesting schedule—Eduardo wouldn't be able to sell his shares anytime soon, so really his ownership was still on paper—like Mark and Dustin and Sean, he assumed. Furthermore, there was also included a general release of any claims against Mark and the company; basically, if Eduardo signed the papers, he'd be saying that these new papers outlined his position at Facebook in its entirety—that everything that came before was simply history.

Sitting there in the dormlike house, listening to the clack of Dustin and Mark's fingers against the computer keys, Eduardo read through the papers again and again. Part of him knew that these papers were important—that they were legal documents, that signing them was a big step forward for the company—but he felt protected, first, because the lawyers were there—Facebook's lawyers, which meant, in his mind, that they were *his* lawyers as well—and more important, because Mark, his friend, was there, Mark was telling him that these documents were necessary and good. Parker was somewhere else in the house—and now, legally, he'd be part of the team for good—but he *had* brought in investor money, and he was one of the smartest people in Silicon Valley.

The important thing was, Eduardo would still have his percentage of the company. Sure, there would be dilution, but wouldn't they all be diluted together? Did it matter that it was no longer thefacebook—wouldn't he be in the same position with Facebook?

He thought back to a few conversations he'd had with Mark recently—about school, about life, about what he should be doing in

Cambridge while Mark was in California. There had been a bit of a miscommunication, in Eduardo's mind—at some points, Mark seemed to be telling him that he didn't need to work that hard for the company while he was in school, that they were going to hire salespeople, that he could step back—and Eduardo, for his part, had maintained that he still had the time to do what was necessary for Facebook.

Well, these papers seemed to say—in Eduardo's mind—that he was just as big a part of the company as he'd ever been. Things might change a bit going forward as more money came in, as more people were hired—but the papers were just a necessary restructuring.

Weren't they?

In any event, Mark had also told him that there was going to be a party, something really cool, when the site reached a million members. Peter Thiel was going to throw it at his restaurant in San Francisco, and Eduardo would have to make the trip back out, because it was going to be well worth the flight.

Thinking about that party, Eduardo had to smile. *Just a necessary restructuring, some legal paperwork that had to be done.* Everything was going to work out just fine. A million members. It was a crazy thought.

He'd definitely come back out to California for that, he thought to himself as he reached for a pen from one of the lawyers and began signing the legal documents. After all, now he owned 34 percent of Facebook—he had reason to celebrate.

Didn't he?

CHAPTER 27 | DECEMBER 3, 2004

Eduardo's eyes burned and his ears rang as he stumbled through the hip and pretty crowd, his head spinning from the music—a throbbing mix of techno, alternative, and rock—and the bright, multicolored lights that swirled across the domed ceiling high above: purples, yellows, oranges, circular patterns twisting and curving like galaxies going supernova, washing the entire restaurant in a truly psychedelic glow.

The place was called Frisson and it was currently the hottest lounge in downtown San Francisco. The decor was somehow exceedingly modern and painfully retro at the same time—landing somewhere between the bridge of the starship *Enterprise* and a 1960s psychedelic drug trip. Eduardo's head was really spinning by the time he'd gotten through the thick of the crowd, partly because of the fairly massive amount of alcohol he'd already consumed, but mostly because he was suffering from major culture shock, having just flown in once again from the staid, and frozen, Harvard campus.

He paused a few yards from the DJ booth that was

planted at the head of the circular dining area, and surveyed the crowd and the posh restaurant. He had to admit, the restaurant was a pretty good choice for Facebook's Millionth Member Party—the shindig Mark had invited him to, set up to celebrate the millionth account activated on the Web site, just days earlier—and barely ten months from the time they'd launched the thing in Mark's Kirkland dorm room. Frisson was modern, hip, and exclusive, just like Facebook. It also happened to be owned by Peter Thiel, who was paying for the party out of his own deep pocket.

Eduardo watched the young, Northern California crowd bouncing to the music; it was almost an even mix of jeans and collared shirts and sleek black European-style duds. Overall, the party was very Silicon Valley, very hip San Francisco. And it was also very Facebook. Much of the room was college-aged, or close to it. Lots of Stanford kids and fresh graduates. Everyone was drinking colorful mixed drinks, and everyone seemed to be having a good time. Eduardo couldn't help noticing the group of cute girls on the other side of the DJ booth. One of them seemed to smile at him, and he blushed, quickly looking away. Yeah, he was still pretty shy, despite everything that had changed in his life.

The party had been going pretty well for him, too. Since he'd walked through the door, he'd been telling everyone who would listen that he had cofounded Facebook along with Mark and Dustin. Sometimes the girls smiled and sometimes they just looked at him like he was crazy. It was a little strange—at Harvard, everyone kind of knew him, what he had done. Here, they were all looking at Mark—and only Mark.

But that was okay, really. Eduardo didn't mind being in the background, here in California. He hadn't gotten into this for the fame. He didn't really care if people knew he had been there in that dorm room, that he owned more than 30 percent of the company, that he was the person most responsible for those million members—other than Mark. He only cared that these people loved the site, and that it was turning into one of the biggest businesses in Internet history.

He grinned sloppily at the thought, then shifted his eyes past the dance floor, to the lounge tables on the other side of the restaurant. Toward the back of the room, seated around a circular table, he could barely make out Mark and Sean and Peter, sitting together, deep in conversation. He knew that coincidentally, it happened to be Sean's birthday—how old was the kid now, twenty-five? He considered heading over to them, but at the moment, he felt a bit more comfortable lost in the crowd, anonymous—alone. The culture shock, again; this place felt so far from Harvard Yard that he might as well have been on the starship *Enterprise*.

He blinked, letting the swirl of lights wash through him.

This place, this restaurant—it was so much to take in. It felt so completely foreign. It felt so—fast. He'd known it from the minute he'd gotten out of the cab in front of the place. Peter Thiel's Ferrari Spyder was parked at the curb outside. Mark's Infiniti—the one he'd been given when his own Craigslist car hadn't been able to get him to that business meeting on time—was somewhere down the street. Maybe next to Parker's BMW.

Eduardo still lived in a dorm room. He walked to classes, through the now snow-covered Yard, lost in the cold shadows of Widener Library.

Okay, he'd been wrong—things had changed pretty dramatically since the beginning of summer. But it was okay. It was a choice he had made. He had nobody to fault but himself. He could have moved out to California. He could have taken time off from school. Anyway, he was a senior, now, only five months to go before graduation. Then he could throw himself into Facebook like the rest, go right back to where he and Mark had started.

For now, tonight, he was going to enjoy himself. He was going to have another drink. He was going to go talk to the pretty girl on the other side of the DJ booth. And then tomorrow, he was going to fly back to Cambridge and get back to his schoolwork. Mark had Facebook under control.

He was pretty sure everything was going to be just fine.

▶ ▶ ▶

Seated at the circular table in the lounge beyond the dance floor, Sean Parker leaned back against a modern Deco chair, listening to Thiel and Mark go on about the new applications they were contemplating for Facebook. Better ways to allow college kids to find one another on the network. Enhancements to the already popular wall where kids could share info. Maybe even a future photo-sharing app—still maybe half a year away—that would rival anything anyone else had come up with. Innovation after innovation after innovation.

Sean smiled to himself; everything was going exactly according to plan. Thiel and Mark were a great match, as he'd suspected.

He took a deep breath, looking beyond his two partners and out into the crowd. Almost immediately he caught sight of Eduardo Saverin, talking to a cute Asian girl by the DJ. Eduardo looked as lanky and awkward as usual, hunched forward as he hit on the girl. She seemed to be smiling, which was good. Eduardo was happy, the girl was happy, everyone seemed happy.

It had all gone so smoothly. Eduardo had signed the necessary legal papers, and had executed the restructuring agreements. Thiel had given them the money they needed to continue flying forward. Facebook had passed a million users, and they were adding tens of thousands more a week. Pretty soon, they'd be opening it up to more schools, more campuses. Eventually, maybe even high schools. And after that—who knows? Maybe Facebook would one day be open to everyone. The college format, the exclusivity—it had already worked its magic. People *trusted* Facebook. People *loved* Facebook.

People were going to want to pay billions for Facebook.

CHAPTER 28 | APRIL 3, 2005

"And there it is. It's official. Spring has come to New England."

Eduardo grinned as his buddy AJ pointed at the girl with the superbly toned legs strolling by the base of the stone library steps, her nose buried in an economics textbook, her flowing blond hair raining down around the wires of her ivory-white iPod.

"Yep," Eduardo responded. "The first short skirt of the season. It's all downhill from here."

Eduardo didn't think he'd ever get used to how long winter seemed to last at Harvard; just a week ago, the Yard was white with snow, these very steps covered in sheets of ice, the air so sharp and cold it hurt to take a breath. It had seemed like March didn't even have a spot on the Harvard calendar—it was just February, February, and more goddamn February.

But finally, finally, the snow was gone. The air smelled alive, the sky was bright and blue and pretty much cloudless, and the girls had started to rearrange their closets, putting the thick, ugly sweaters away and reaching for the skirts, the cute little tops, the open-toed shoes. Well, maybe the tops weren't all that cute—it was Harvard, after all—but the skin was showing, and that was a damn good thing.

Of course, it could change on a dime; tomorrow, those gray clouds could roll back in and the Yard could turn back into an inhospitable lunar landscape. But then again, tomorrow, Eduardo wouldn't be in New England. He'd be back in California once more, because he'd been summoned from on high.

AJ gave him a wave, then headed down the stone steps, on his way to a seminar on the other side of the Yard. Eduardo would follow in a few minutes—but he wasn't in any rush. They were seniors, barely two months from graduation. They could be late to class. Hell, they could skip class entirely, it wouldn't make any difference. As long as they passed the few exams they had left, they were pretty much on their way out of Harvard, with those golden diplomas that supposedly meant so much in the real world.

The real world. Eduardo wasn't even sure what that term meant anymore. It certainly wasn't California, where Mark was still holed up, in yet another sublet in another leafy suburban town, furiously building Facebook ten thousand users at a time. And it wasn't the new Facebook offices in Palo Alto that Mark had told him about, the ones they were putting the finishing touches on before the upcoming round of new hiring—the growth they'd talked about back in the fall, when they'd signed all the papers for the company restructure.

The real world couldn't have anything to do with Facebook, because the real world simply didn't move that fast.

One million members had suddenly become two, on its way to three. And the little Harvard-based Web site was now simply everywhere—on five hundred campuses, in every newspaper Eduardo saw at the newsstand, on every news show he happened to catch before or after classes. Everyone he knew was on Facebook. Even his dad had logged on, using his account, and had loved what he'd seen. Facebook wasn't the real world—it was way bigger than that. It was a whole new universe, and Eduardo couldn't help but be proud of what he and Mark had done.

Even though, over the past two months, he'd had almost no significant interaction with the guys out in California—other than the odd phone call, the odd request for a contact from New York or a name from his research into potential advertisers. In fact, Eduardo had been so distant from Mark over the past couple of months, Eduardo'd had time to launch a whole separate Web site—something called Joboozle that aimed to be a sort of Facebook for jobs, where kids could go to search out potential employers, share résumés, network. Eduardo didn't have any expectations that Joboozle would ever be anything close to Facebook, but it certainly had passed the time while he waited for Mark to get back in touch.

And finally, Mark *had* gotten back in touch—an e-mail, just a couple days ago, asking him to make the trip back out to Cali. Something about an important business meeting, and a new hire that Eduardo was supposed to help train.

In the e-mail, Mark had also mentioned something that had caused Eduardo a little bit of concern. Recently, some big-name venture capital funds had been circling the company—Sequoia Capital, the biggest fund in Silicon Valley, run by Sean Parker's old nemesis Michael Moritz, and Accel Partners, a very prestigious Palo Alto fund that had been active in the space over the past decade, and Mark had intimated in the e-mail that there was a chance they'd let one of the funds invest. Mark had also mentioned that Don Graham, the CEO of the Washington Post Company, was interested as well.

Furthermore, Mark had noted, he and Sean Parker and Dustin were thinking about selling a little of their own stock if a deal went through—two million dollars apiece was the figure he gave in the e-mail.

Eduardo had been more than a little surprised by that; first, from the papers he'd signed, he was pretty certain that he didn't have the ability to sell stock—his shares didn't vest for a long, long time. So why were Mark, Sean, and Dustin able to cash out two million dollars' worth? Hadn't they signed the same papers as he had, during the restructure?

And second, why was Mark talking about selling shares at all? Since when did Mark care about money? And why did Sean Parker get to make two million bucks when he'd been a part of the company officially for about ten weeks? Eduardo had been there since the beginning.

It certainly didn't seem fair.

Maybe Eduardo was simply misunderstanding the situation. Maybe Mark would clear things up when Eduardo met with him in California. In any event, Eduardo had decided he wasn't going to let his emotions take over this time—since his anger hadn't exactly helped the situation back during the summer. He was going to be calm, rational, and understanding. It was spring, the skirts were out, and school was almost over.

Tomorrow, Eduardo would make the six-hour trip, check out the new offices that were under construction, attend that business meeting, and train that new hire, whoever he was. Hopefully, it would be the beginning of things going back to normal between him and Mark—so that when he graduated, he could go right back to his old role as Mark's founding partner. The idea was pretty pleasing to him—because in a way, it meant he could extend his college life even further, because as big a company as Facebook became, Eduardo was pretty sure it would always feel like college to him. At Facebook, he could keep on postponing the real world, just like Mark was doing, maybe forever.

Eduardo was warmed by that thought as he started down the library steps toward the Yard. Tomorrow, he'd be back with Mark—and Mark would explain everything.

CHAPTER 29 | APRIL 4, 2005

Eduardo would remember the moment for the rest of his life.

He started to shake as he stood there in the mostly bare office, staring down at the papers that the lawyer had handed him the minute he'd walked through the door. It was a different lawyer, this time, and it was a different door; not the dormlike sublet in a leafy suburb, but a real office, on University Avenue in downtown Palo Alto, with glass walls, maple-covered desks, new computer monitors, carpeting, even a staircase covered in graffiti by a local artist who'd been commissioned for the task. A real office, and another real lawyer—standing between Eduardo and Mark, who was somewhere inside, at one of the computers, where he always seemed to be, safe in the glow of that goddamn screen.

At first, Eduardo had thought the guy was joking, greeting him with more contracts to sign, even before he'd had a chance to check out the place, or ask Mark about the new hire, the two-million-dollar stock sale, the e-mail. But as Eduardo started to read the legalese, he'd realized that this trip to California wasn't about a business meeting.

This was an ambush.

It took Eduardo a few minutes to understand what he was reading—but as he did, his cheeks turned white, his skin going cold. Then full realization hit him like a gunshot to the chest, shattering him from the inside out, destroying a part of him that he knew he'd never get back. No amount of hyperbole, no adjectives, no words—nothing could describe what it felt like—because even though, deep down, he should have seen it coming, he should have known, goddamn it, he should have seen the signs—he simply hadn't. He'd been so fucking blind. *So fucking stupid.*

He simply hadn't expected it from Mark, from his friend, from the kid he'd met when they were two geeks in an underground Jewish fraternity trying to fit in at Harvard. They'd had their problems, and Mark had the ability to be pretty cold, pretty distant—but this was way beyond that.

To Eduardo, this was a betrayal, pure and simple. Mark had betrayed him, destroyed him, taking it all away. It was all right there, in the papers in his hands, as clear as the pitch-black letters imprinted on those ivory-white pages.

First, there was a document dated January 14, 2005—a written consent of the stockholders of TheFacebook to increase the number of shares the company was authorized to issue up to 19 million common shares. Then, there was a second action dated March 28, issuing up to 20,890,000 shares. And then there was a document allowing the issuance of 3.3 million additional shares to Mark Zuckerberg; 2 million additional shares to Dustin Moskovitz; and over 2 million additional shares to Sean Parker.

Eduardo stared at the numbers, rapidly doing the calculations in his head. With all the new shares, his ownership of Facebook was no longer anywhere near 34 percent. If just the new shares had been issued to Mark, Sean, and Dustin, he was down to well below 10 percent—and if all the authorized new shares were issued, he'd be diluted down to almost nothing.

They were diluting him out of the company.

The lawyer started to talk as Eduardo looked at the papers. Eduardo wondered what Mark would expect him to do. Or maybe Mark didn't think Eduardo was going to have any reaction at all. Maybe Mark believed that Eduardo had already left the company a long time ago—back in the fall, when he'd signed the papers that had made all this possible. Or maybe even earlier than that, during the summer, after he'd frozen the bank accounts. Two different wavelengths, two different points of view.

The lawyer droned on, explaining that the new shares were necessary, that there were interested VCs who would need them, that Eduardo's signature was a formality, that the shares had already been authorized anyway, that it was good and necessary for the company, that it was a decision that had already been made—

"No."

Eduardo heard his own voice reverberate through his head, bounce off the glass walls, up the graffiti-marked staircase, throughout the near-empty office.

"No!"

He refused to sign away his ownership of Facebook. He refused to sign away his accomplishment. He had been there in the beginning. He had been in that dorm room. He was a founder of Facebook and he deserved his 30 percent. He and Mark had an agreement.

The lawyer's response was immediate.

Eduardo was no longer a member of Facebook. He was no longer part of the management team, no longer an employee—no longer connected in any way. He would be expunged from the corporate history.

To Mark Zuckerberg and Facebook, Eduardo Saverin no longer existed.

Eduardo felt the walls closing in around him.

He had to get out of there.

Back to Harvard. Back to the campus, back home.

He could not believe what he was hearing. He could not believe the betrayal. But he had no choice, he was told. The decision had been made, he was told—made by Mark Zuckerberg, the founder and CEO, and by the new president of Facebook.

Eduardo had one more thought as the horrible news washed over him.

Who the hell was the new president of Facebook?

When he thought about it, he realized he already knew the answer.

CHAPTER 30 | WHAT GOES AROUND . . .

Sean Parker hit the sidewalk soles first, launching himself out of the BMW with a burst of pure, frenetic energy. His brain was moving at ten thousand rpm's, even faster than usual, because he was, metaphorically, on his way to the sweetest dessert of his life.

He slammed the car door shut behind him, then stepped to one side, leaning back, arms crossed against his chest. He looked up at the glass-and-chrome building that housed Sequoia Capital's main offices. *God, how he hated this place.* He remembered, with more than a little irony, how different he had once felt—how he'd once come here, looking for funding, for a partnership, for attention, for anything. How he'd gotten that attention—and had ended up out on his ass, pushed out of the company he'd started himself, that he'd built with his own sweat and tears.

How different things were now. This time, it was Sequoia doing the begging. Call after call, they'd hounded the Facebook offices, trying to set up a meeting, trying to get Mark on the phone, trying to get him into a room for a pitch. Hell, everyone was calling now, all the big names. Greylock, Merritech, Bessemer, Strong, everyone. And not just the VCs. There were already rumors growing that Microsoft and Yahoo were watching. And Friendster had already made an informal offer; ten million—chickenshit money—which Sean and Mark had easily turned down. MySpace was interested as well—hell, everyone wanted in now. And Sequoia, the biggest boy on the block, certainly didn't want to be left out in the cold.

So Sean had stalled them awhile, picturing Moritz stewing in his secluded lair, shouting at his peons in that bizarre, villainous Welsh accent. Sean guessed that by now, Moritz must have known that he was behind Facebook's reticence to meet and greet; but in Sean's view, the megalomaniac probably thought Sean would give in sooner or later. And just when they were frothing at the mouth, Sean had seemingly done exactly that, setting up this morning's meeting.

Now here he was, grinning like a crazed monkey. He was dressed all in black, like the car, from his thin DKNY pants to his crocodile belt. Batman, out for justice, hitting the streets of downtown San Francisco to set things right again.

He heard the driver's-side door slam shut, and turned to see Mark coming around the front of the car.

"Sweet Jesus," Sean murmured—and his grin turned into a full-throttled laugh.

Mark was dressed in brightly colored pajamas, his laptop under his arm. His hair was a complete mess, but there was a serious look on his face.

"You sure about this?"

Sean laughed even louder. Oh yes, he was more certain about this than anything he'd ever done before.

"It's perfect."

Then Sean glanced at his watch. Really, perfect.

Not only was Mark showing up ten minutes late to a meeting with the biggest venture capital firm in Silicon Valley, but he was going to walk in there like the craziest motherfucker in town. Sean wasn't going to go to the meeting—that would have simply been too much, even for him—but Mark would be able to handle himself just fine. Mark was going to apologize, tell 'em he had overslept, and hadn't even had time to get dressed. Then he was going to launch right into his pitch. When he was nearly finished, he'd open up the PowerPoint that they had concocted especially for the Sequoia boys—and what was on the PowerPoint was going to twist the knife in even deeper. And then Mark was going to walk right out of there.

Seqouia Capital would never—*never*—have the opportunity to invest in Facebook. Sean would make sure of that. Mark had seen exactly what Moritz and the Sequoia boys had done to him, kicking him out of Plaxo, cutting him off at the throat. And Thiel was in utter agreement—because Sequoia had treated him badly during the PayPal days as well. Sequoia would learn the ultimate lesson of this small town: what goes around comes around.

And Mark and Sean wouldn't feel a thing, because everyone wanted a piece of Facebook, now. Sure, they'd turned down Friendster—but there was one deal waiting in the wings that they both knew they were going to accept. Accel Partners, one of the most prestigious VC firms around, had been chasing them for weeks. Whenever Jim Breyer, Accel's leading partner, one of the most brilliant VCs in the business, called, Sean had grabbed the phone and screamed crazy numbers at him. One-hundred-million valuation or nothing! Two hundred million or bust! And Breyer had finally gotten the picture.

Simultaneously, Mark had also been talking a lot with Don Graham, the head of the Washington Post Company, a man that had become somewhat of a friend and mentor to Mark; it was an interesting pairing, an interesting idea—that of a media titan with the genius behind a social revolution built on the sharing of information. Mark was considering doing a deal with Graham and the *Washington Post*—which had pushed Accel to get even more serious, and the wind was beginning to blow clearly.

Very soon, Accel was going to invest close to thirteen million for a small stake in the company—an investment that would put Facebook's valuation at close to one hundred million dollars. After only fourteen months. *One hundred million.* And that, too, was just a starting place. Within six months, Sean was certain they would triple that valuation. By the end of 2005? Who knew where they could be? If people continued to sign up at the current rates, they'd be at fifty million users within a year.

Sean had a pretty good feeling that his billion-dollar baby was about to be birthed.

He grinned as Mark walked past him, heading slowly toward the Sequoia building. Part of him wished he could attend the meeting with Mark—but it was good enough, just picturing it in his mind as it took place. He gave Mark a final wave of encouragement.

"This is going to be great."

Then Sean took one more look at those pajamas—and laughed out loud.

This was going to be fucking awesome.

CHAPTER 31 | JUNE 2005

"Ten thousand men of Harvard . . ."

Eduardo's knees cried out as he twisted his lanky body beneath the heavy folds of the black polyester gown, trying to find a comfortable position against the little wooden folding chair beneath him, trying to somehow fit his long frame into that tiny space, jammed as he was between similar chairs on all four sides. It was ridiculously hot beneath the gown, and it didn't help that the stupid square hat on his head was at least two sizes too small, pinching at the damp skin of his forehead and yanking strands of his hair out by the roots.

Even so, Eduardo felt himself smiling. Even after everything that had happened, he was smiling. He looked to his right, down the long row of his classmates in their matching, jet-black gowns and silly hats. Then over his shoulder—at the row upon row upon row of similarly attired seniors, stretching halfway back across the Yard, right up to where the black gowns gave way to light summer blazers and khaki pants, to the colorful sea of proud families with their cameras and their digital video recorders.

"Ten thousand men of Harvard . . ."

Eduardo turned back toward the stage, which was a good ten yards ahead of him. President Summers was already behind the podium, flanked by his deans, a huge bin of diplomas to his right. Any minute now, the microphone on the lectern in front of the president would burst to life, and the first name would echo through the Yard, bouncing off the ancient brick buildings covered in ivy, reverberating over the stone steps of Widener, rappelling up the library's great Greek pillars, up into the aquamarine sky.

It had been a long morning already, but Eduardo was filled with energy—and he could tell that his fellow seniors felt equally alive, fidgeting anxiously against the little wooden seats.

The day had begun early, with the march from the River Houses—the long line of seniors garbed in black gowns traipsing through Harvard Square and down into the Yard. Although it was hot outside, Eduardo had his jacket and tie on under the gown. After the ceremony, he was going to spend most of the afternoon with his family. He wasn't quite sure where they were in the gathered audience that stretched out behind where the seniors were sitting, but he knew they were there.

In truth, the entire Yard was packed with people—more people than Eduardo had ever seen in one place, outside of the odd rock concert he'd gone to in high school. And they'd be there all day. Later that afternoon, John Lithgow, the actor and Harvard grad, would be speaking. Before that, the graduating seniors would gather on the steps of Widener for a class photo. They'd go to a picnic with their families, and then they'd say good-bye to one another and to the school. Maybe some of them would throw their square hats into the air—because they'd seen the clichéd act on television, and well, the hats were pretty stupid anyway.

Eduardo turned his attention back to the stage. He was immediately impressed by all the color, the stark contrast to the sea of black that surrounded him. The university marshals, the tenured professors, the honored alum—they were all present now, lined up behind the president in their bright, nearly psychedelic gowns. Eduardo's gaze slid back to that bin of diplomas. He knew that somewhere in that mountain of rolled paper sat a diploma with his name on it; a curled, Latin-embossed page that had cost his parents more than a hundred and twenty thousand dollars.

In some ways, that diploma had cost Eduardo much, much more.

"Ten thousand men of Harvard . . ."

The melody was coming somewhere from Eduardo's left. He couldn't believe that someone actually knew the words to the old college fight song. Well, some of them anyway—whoever it was, the guy was humming his way through most of the tune. Eduardo *did* actually know the words, because he'd learned them his freshman year after the marching band had sung the song during the Harvard-Yale game. He'd been pretty gung ho "Crimson" at the time, so proud that he was a part of this history, this university. So proud, because his father was so proud, because all the hard work of high school had paid off. The difficult road—learning a new language, fitting into a new culture—had led to this place, this beautiful Yard embraced by these historic buildings. He had learned the song because this was his moment, as much as it belonged to anyone who'd ever stood shoulder to shoulder in this place. He'd earned it, every second of it.

Ten thousand men of Harvard want vict'ry today,
For they know that o'er old Eli
Fair Harvard holds sway.
So then we'll conquer old Eli's men,

And when the game ends, we'll sing again:
Ten thousand men of Harvard gained vict'ry today!

He turned his attention back to the stage. Summers was almost ready behind the lectern, his wide, jowly face just inches from the microphone. Eduardo knew it would take them a while to get to his name, and when they did, he also knew that the president would probably mispronounce it. Leave the *O* off the first part, or lean heavily on the second syllable of the last. He was used to that, and he didn't care. He was going to march up there and get that diploma, because he deserved it. That was how the world was supposed to work. That was fair.

Just as the microphone burst to life and the first name was read, a flash went off from somewhere behind Eduardo, a high-powered camera catching the first senior on his way to the stage.

Eduardo couldn't help wonder if that picture would one day find its way onto someone's Facebook profile. He was pretty certain that, sooner or later, it would.

For the first time that day, his smile almost disappeared.

▶ ▶ ▶

Two A.M.

Eighteen long hours later.

Hands jammed deep into the pockets of his blazer, head swimming from a day of family, scorching temperatures, and a quarter bottle of expensive Scotch, Eduardo sank deep into a leather couch on the third floor of the Phoenix, watching a group of blond girls he didn't know dancing around a coffee table piled so high with alcohol bottles, it looked like a little glass metropolis, sparkling brightly on a moonlit night.

Downstairs, the party was in full swing. The entire three-story building was throbbing from the music coming from the dance floor on the first floor, a mix of hip-hop and Top 40; Eduardo could picture the surging crowd of kids trampling the hardwood floors, inhaling the smoke from the bonfire outside, kicking up the dander of two hundred years of history as they bucked and spun to the beat. He could picture all the pretty girls, many of them still fresh from the Fuck Truck, and all the eager young Phoenix members, searching for that special connection, that night to remember, that frozen moment in time.

But up here, on the third floor, things were quieter. Aside from the dancing blondes, the place had the feel of a posh VIP room. And the decor was pure VIP as well: plush crimson carpeting, deep, wood tones on the walls and ceiling, the leather couches, the tables teeming with expensive brand-name bottles of liquor. This third-floor parlor was utterly exclusive, invite only, totally velvet rope.

Since Eduardo had returned from California—since the moment he now mostly referred to as Mark's betrayal—he'd spent a lot of time in this room, sitting on this couch. Thinking. Contemplating. Planning out his future.

College was over, now, and Eduardo was heading out of the safe confines of the Yard. He wasn't sure where, yet—maybe Boston, maybe New York. But he did know that he wasn't a kid anymore. He didn't feel like a kid anymore.

For one thing, he'd already begun the legal process of going after what he felt was fairly his. He'd hired lawyers, sent out letters, made clear his intentions to Mark and the rest of the Facebook team—he intended to sue. He hated the idea of a courtroom, of going up against his "friend" in front of a judge or a jury. But he knew that there was no other way. It wasn't just Mark and him anymore.

Sitting there on the leather couch, he wondered if Mark had any regrets at all at how things had turned out.

Probably not, he realized with a grimace. Mark probably didn't even think that he'd done anything wrong. From Mark's point of view, he had only done what was necessary for the business.

Facebook had been Mark's idea in the beginning, after all. *He* was the one who'd put in the hours, put in the work. *He'd* built the company from the dorm room up. *He'd* written the code, launched the site, gone to California, postponed college, found the funding. To him, it had been a Mark Zuckerberg production from day one. And everyone else was just trying to hang on. The Winklevosses. Eduardo. Maybe even Sean Parker.

In fact, from Mark's point of view, it was probably Eduardo who had acted inappropriately, who had betrayed their friendship. From Mark's point of view, Eduardo had tried to hurt the company by freezing the bank account. From Mark's point of view, Eduardo had tried to make it difficult to raise VC money by asserting his own position as the titular head of business. From Mark's point of view, Eduardo had even done some other things that could have caused Facebook harm, such as starting a separate Web site, Joboozle, and approaching the same potential advertising base with what Mark might have seen as Facebook's trade secrets. Mark had as much reason to see himself as the wronged party as Eduardo did.

But Eduardo didn't see it that way. He believed, fully and completely, that he had been there from the beginning. That he had been integral to Facebook's success. He had put up the initial money. He had put in his time. And he deserved what they had agreed upon. Pure and simple.

He did agree with Mark about one thing—it wasn't about friendship, anymore. It was business. Simply business.

Eduardo would pursue what he believed he deserved. He'd take Mark to court. Make him explain himself. Make him do what was fair.

As he watched the girls gyrate to the music, their blond hair flowing and twisting above them in a swirling, golden storm, he wondered if Mark even remembered how it had all started. How they had been two geeky kids trying to do something special, trying to get noticed—really, trying to get laid. He wondered if Mark realized how much things had changed.

Or maybe Mark had never really changed at all; maybe Eduardo had just misread him from the start. Like the Winklevoss twins, Eduardo had projected his own thoughts onto that blankness, drawing in the features he most wanted to see.

Maybe he'd never really known Mark Zuckerberg.

He wondered if, deep down, Mark Zuckerberg even knew himself.

And Sean Parker? Sean Parker probably thought he knew Mark Zuckerberg, too. But Eduardo was pretty sure that was going to be a short-lived pairing as well.

In Eduardo's mind, Sean Parker was like a jittery little comet tearing through the atmosphere; he'd already burned through two startups. The question wasn't *if* he'd burn through Facebook as well, it was *when.*

CHAPTER 32 | THREE MONTHS LATER

The strange thing was, nobody even heard the sirens.

One minute, everything was going along great. The party was really rocking, the suburban house filled with good-looking, happy people. College girls and grad-student guys, urban hipsters and stylish twentysomethings, kids with back-packs and baseball hats mingling with professionals in tight-fitting jeans and collared shirts; the place felt like an extension of any cosmopolitan nightclub scene, but in a manageable, collegiate setting—kind of like a frat party for kids who didn't know the first thing about frats. The booze was flowing, the music pounding through the wood floors and reverberating off the bare plaster walls—

And then, *blam,* in the blink of an eye it all went bad.

There was a scream, and then the front door crashed open. Flashlights tore across the dark, crowded dance floor, darting and diving along the plaster walls like UFOs assault-ing a barren plain. And then they came pouring in, like so many fucking gestapo bullyboys, shouting and barking and shoving, wielding those flashlights like goddamn light-sabers.

Dark blue uniforms. Drawn nightsticks, and badges, and even a few handcuffs. No guns that anyone could see, but the holsters were clearly visible, the cruel twists of metal bulging through the thick dark rubber sleeves.

Sirens or no, this party was *over.*

One can imagine that Sean Parker's first thought was that someone had made a mistake. This was just a goddamn party, right outside a college campus. It was totally innocuous. He'd gone there with one of Facebook's many undergraduate employees, a pretty girl whom he'd befriended—pure, innocent fun. Just a party, the kind of thing he'd been to a thousand times before. Utterly harmless, nothing crazy going on at all.

Well, okay, maybe there *was* alcohol in the house. And maybe the music was a little bit too loud. And, sure, maybe *some* of the kids had been doing a little coke, smoking a little pot. Sean didn't really know—he hadn't spent much time in the bathroom since he'd arrived at the house, he'd been busy on the dance floor. Other than the inhaler in his pants pocket and the EpiPen full of epinephrine in his shirt, he was as clean as the pope. His chronic asthma and ridiculous fucking allergies made certain of that.

Who cared, anyway? It was a *party.* There were a lot of college kids present. Wasn't college supposed to be about experimentation?

Revolution?

Freedom?

Shouldn't the cops have been a little more forgiving, considering the locale?

But the looks on the cops' faces were anything but forgiving. No question about it, Batman was in for a hell of a fucking night.

It dawned on him, then, that maybe this wasn't as much about bad luck—about being in the wrong place at the wrong time—as it was about being Sean Parker in the wrong place, at the wrong time. Maybe, just maybe, this wasn't as simple as a party that had gotten too loud. Maybe, once again, he'd become a target.

Facebook wasn't a little dorm-room company anymore; Sean had seen to that himself. It was now a major corporation, on its way to a billion-dollar valuation. And he and Mark, they weren't two kids playing around with a computer program, they were executives running a company—a company that neither one of them wanted to sell, a company that both of them now believed would one day be worth much, much more than a billion dollars.

The growth that had gone on over the past few months was nothing less than spectacular. In Sean's view, what was going on with Facebook was truly transformative, the culmination of a few brilliant ideas played out across an exceedingly successful network of eager participants.

The first, and most recent, transformative development had to be the picture-sharing application, the idea that Facebook was now a place where you shared and viewed pictures that coincided with your social life. It was the true digitalization of real life: you didn't just go to a party anymore, you went to a party with your digital camera so you and your friends could relive that party the next day—or at two in the morning—via Facebook. And the tagging, the idea that you could tag anyone you wanted in those pictures, so that those people could find themselves, see who was there, literally see your social network in its digital form—it was utter genius. And it had led to an explosion of users—now maybe eight million, ten million, God, Facebook was growing so fast.

And they weren't even close to finished: the next transformative step on par with pictures would be the newsfeed, an idea that Sean and Mark had been thinking about independently. The newsfeed would be a constant updating of information among people in a social network, which would link people even more through their Facebook pages—a living, digital log of every change in a person's profile broadcast to all his friends instantaneously. When completed, it would be a sophisticated feat of computer engineering that Dustin and Mark would have to pull off—exponentially complex, a sort of broadcast channel limited to groups of friends that had to be constantly updated, moment by moment. For Sean, the idea had come about after hours spent watching what people did when they logged into Facebook; how they always checked their friends' status updates, checked to see which friends had changed their profiles, their photos. The idea of a newsfeed was one of those eureka moments—if there was a way this could happen automatically, Sean had realized, it would enhance the Facebook experience the same way photos and tagging had.

These were more than just applications—they were milestones in the making, changing what began as a dorm-room idea into a life-changing, billion-dollar company. Building the biggest, most successful picture-sharing site on the Web *on top* of the most successful social network? Adding an innovation like a newsfeed on top of that?

Facebook was going to be bigger than anything else on the Web, Sean was sure of it. Someday soon they'd open it up to the general public—the next, great transformative step, the next milestone—and then they'd go international. And after that, well, nothing would ever come close to Facebook again. Sean wasn't thinking Friendster or even MySpace: he was thinking Google and Microsoft.

Facebook would be *that* big.

And when things got big—well, Sean Parker knew better than

anyone else what often happened. People began to act differently. Friendships fell apart. Problems arose—sometimes seemingly out of nowhere.

Maybe, just maybe, as Facebook got bigger than big, as the money poured in and the VCs started to think in terms of billions—maybe there were people who didn't feel they needed a Sean Parker involved anymore.

It had happened before—twice. Could it really be happening again?

Or was he just being paranoid? Maybe things were exactly as they seemed. A party that was being busted—and him right there in the middle of it all.

Bad luck.

Bad timing.

Sean's next thought, as he was arrested, was that he had to make a phone call. Speculation was a beast that could cause a lot more damage than a nightstick or a pair of handcuffs. Innocent or not, it didn't exactly look good for the president of a transformative, world-changing, billion-dollar company to get busted with an undergraduate employee at a house party. He didn't think he was going to end up in jail—but he was certain of one thing:

Innocent or not, setup or pure bad luck, Mark Zuckerberg was going to be pretty pissed off.

CHAPTER 33 | CEO

At some point that night, or maybe even the next day, Mark Zuckerberg likely received a phone call; maybe from the corporate lawyers, maybe from Sean himself. The odds are good that Mark was at the Facebook offices at the time—because he was almost always at those offices. We can picture him there, alone, his face lit by the greenish-blue glow of the computer screen on the desk in front of him. Maybe it was still the middle of the night, or maybe early morning; time had never been a very useful concept to Mark, just twitches in a clock that had no real-world purpose, no claim or innate value. Information was so much more important, and the information Mark had just received certainly had to be dealt with quickly—and with utter efficiency.

Sean Parker was a genius, and he'd been instrumental in getting Facebook to where it was now. Sean Parker was one of Mark's heroes, and would always be a mentor, an adviser, and maybe even a friend.

But we can imagine what Mark must have thought after

hearing the details of the house party that had just been busted by the cops: Sean Parker had to go.

Whatever the reason, even though Sean wasn't going to be tried or indicted for anything that he'd done—in some people's eyes, the current situation would make Sean a danger to Facebook. To his detractors, he had always been unpredictable, wild—people didn't always understand him, and some found his energy level terrifying. But this was different. This was black and white. No matter why it had happened—whether it was bad luck, or something else—the result was as clear as data in, data out.

Sean Parker had to go.

Like Eduardo, like the Winklevosses, anything that became a threat—no matter the intention—had to be dealt with, because in the end, the only thing that mattered was Facebook. It was Mark Zuckerberg's creation, his baby, and it had become the focus of his life. In the beginning, maybe it had simply been something fun, something interesting. Another game, a toy, like the version of Risk he'd built in high school, or Facemash, the stunt that had nearly gotten him kicked out of Harvard.

But now, we can surmise, Facebook was an extension of the only true love of Mark's world—the computer, that glowing screen in front of his face. And like the personal computer that Mark's idol Bill Gates had unleashed on humanity by means of his groundbreaking software, Facebook really was a revolution—world changing, creating a free exchange of information across social networks that would digitize the world in a way nothing else could.

Mark wouldn't let anything, or anyone, stand in the way of Facebook.

What Mark Zuckerberg had become could best be illustrated by the business card, simple and elegant, with a single sentence printed across the center, that he created, most likely sitting at his computer, the screen glowing across his face; the business card that he would have printed out to carry with him everywhere.

In one sense, the card represented nothing more than Mark Zuckerberg's personal brand of humor. But in another sense, the card was more than a joke—because it was true. No matter what else anyone wanted to believe, no matter what anyone else ever tried to do, the sentiment of the card would always be true.

Inevitably, indelibly true.

We can picture Mark reading the words on the card aloud to himself, the smallest hint of a grin twitching across his usually impassive face.

"I'm CEO—Bitch."

CHAPTER 34 | MAY 2008

Shit, it was going to be one of those nights.

Eduardo wasn't exactly sure what the name of the club was, or even how, exactly, he'd gotten there. He knew it was New York, and he was in the meat-packing district. He knew there had been a cab involved, and at least two friends from college, and somewhere along the line there had been a girl, Christ, there always seemed to be a girl involved, didn't there? And he was pretty sure she was hot, possibly Asian, and she might even have kissed him.

But somewhere between the cab and the club she'd disappeared, and now he was alone, sprawled out on a bright blue leather banquette, staring at his own reflection in a Scotch glass, seeing his own face melting down the curved slopes of the ice inside, like an image from a funhouse mirror, or maybe one of those Salvador Dalí paintings, the ones they'd talked about in that Core class—Spots and Dots, he thought they called it, modern art for kids who didn't really give a shit about modern art.

He was alone, and he was drunk—but really, not *that*

drunk. It was a combination of things that was blurring his vision, and the alcohol was not even that high on the list. First, there was the lack of sleep. It had been about three weeks since he'd gotten to bed before four; with the new start-up he'd been working on—involving health care, social networks, and everything in between—and the lawsuit that dominated many of his days, and of course his social life—spread out between Boston and New York and sometimes California—and the Phoenix, always the Phoenix. Nobody cared that he was a little bit older than everyone else at the club, because they were still brothers, they would always be brothers. And everyone at the Phoenix still knew exactly who he was. *What he'd done.* Even if the rest of the world had never heard of him. Even if the rest of the world only equated Facebook with one name, one kid genius.

Yeah, Eduardo was tired. He hadn't really slept in weeks. He leaned back against the banquette, stared into that Scotch glass—when a sudden memory flashed across his thoughts.

A memory from a night just like this one, another moment when he hadn't kept his mouth shut—a moment from that summer he'd spent in New York, way back in 2004. Eduardo wasn't sure exactly of the day and month, but it had been sometime after he'd frozen that bank account, sometime after those phone calls between him and Mark that had, in retrospect, been the beginning of the end, the cracks that eventually turned into compound fractures. Eduardo had been angry, and he'd been hurt—and he'd gone out drinking, just like tonight, and had ended up in a club, just like this one.

That night, he'd been on the dance floor, chasing after some girl, when he'd glanced across the club, and had noticed someone standing at the edge of the room, looking in his direction.

Eduardo had recognized the kid immediately—because, well, he had been hard to miss. Big, muscular, an athlete with a movie-star face and an Olympic physique. Eduardo had seen him many times around campus, with his identical twin brother. In fact, Eduardo wasn't even sure which of the Winklevoss twins he was looking at. Just that it was one of them, right in front of him, barely ten feet away in some nameless New York club.

Right there and then, Eduardo had let the emotions and the alcohol get the better of him. Maybe, deep down, he'd had a premonition about what was going to happen between him and Mark. Or maybe he had just been drunk.

Whatever the reason, he'd walked right up to the Winklevoss twin, and had held out his hand.

As the stunned kid had stared at him, Eduardo had let the words come spilling out:

"I'm sorry. He screwed me like he screwed you guys."

And without another word, he had turned—and disappeared back onto the dance floor.

EPILOGUE | WHERE ARE THEY NOW . . . ?

SEAN PARKER ▸ After leaving Facebook, Sean Parker has remained a force in the Silicon Valley community; recently, he has been made a managing partner at the Founders Fund, a venture capital firm created by Peter Thiel that focuses on early-stage investments in tech companies, searching out deals similar to the five-hundred-thousand-dollar investment Thiel made in the early days of Facebook's growth, an investment that is now valued at over a billion dollars. More recently, Sean has founded yet another company, the mysteriously titled "Project Agape," a social network aimed at assisting large-scale political activism over the Internet.

TYLER AND CAMERON WINKLEVOSS ▸ Since the end of 2004, Tyler and Cameron Winklevoss have doggedly pursued their legal case against Mark Zuckerberg and Facebook, finally resulting in a settlement late last summer. Although the details of the settlement were sealed by orders of the judge, in recent months leaked information from the law firm that represented the Winklevosses and ConnectU described the terms of the settlement, maintaining a payout in the vicinity of sixty-five million dollars. Though the sum seems significant, there is much evidence that Tyler and Cameron were not happy with the results of the settlement, and it's likely that their battle with Mark and Facebook is far from over. On a brighter note, Tyler and Cameron did make the U.S. Olympic rowing team and competed together in the 2008 Beijing Olympics, placing sixth in the men's pair competition. Since then, they have continued their training, and are currently deciding whether or not to compete again in London in 2012.

EDUARDO SAVERIN ▸ Eduardo Saverin continues to split his time between Boston and New York, and remains a frequent visitor to the hallowed upper floors of the Phoenix. The details of his lawsuit against Mark Zuckerberg and Facebook, and that brought by Mark against Eduardo, have remained shrouded in secrecy; however, in January of this year, Eduardo's name and title as "cofounder" were abruptly reinstated into the Facebook manifest, his very existence reinstalled into the company's corporate history. This development can only be seen as evidence that Eduardo has found some success in his quest to receive credit for his role in the creation of Facebook. Legal issues aside, whether Eduardo and Mark can ever repair their friendship remains to be seen.

FACEBOOK AND MARK ZUCKERBERG ▸ As for Facebook itself, in October of 2007, after a brief and highly public bidding war with Google, Microsoft bought a 1.6 percent stake in the company for 240 million dollars, roughly valuating Facebook at over 15 billion, or more than one hundred times its 150 million dollars in annual revenues. Since then, like the economy itself, Facebook has deflated somewhat in terms of overall valuation while its revenues have continued to increase; but whatever the true multiple might be, Facebook has continued its almost startling pattern of growth. By the end of this year, Facebook's membership will be well over two hundred million users, and according to recent reports, the company is gaining around five million users a week. Highly publicized missteps, such as near debacles involving issues of the ownership of user content and the misuse of "private information" for advertising purposes, have not slowed the social revolution at all, and it seems very likely that Facebook will continue to enhance the lives of an enormous number of people for years to come. Mark Zuckerberg's little dorm-room production has grown into one of the most influential companies on the Internet; and though it's unclear how much Mark Zuckerberg is actually worth today, he is certainly one of the richest twenty-five-years-olds on the planet—and has been described as the youngest self-made billionaire of all time.

ACKNOWLEDGMENTS

This book began—as these things often do—with an e-mail that came to me, completely out of the blue, at two in the morning; I am indebted to Will McMullen for taking that first step, and for introducing me to this story as only he could. My deepest thanks also to Daryk Pengelly, Alasdair McLean-Foreman, and everyone else at Harvard and the Phoenix-S K who aided me in my research into the world behind those ivy-covered gates.

I am immensely grateful to Bill Thomas, my fantastic editor, and his entire team at Doubleday/Random House. I am also indebted to Eric Simonoff and Matthew Snyder, agents extraordinaire. Many thanks to my brothers in Hollywood, Dana Brunetti and Kevin Spacey, and to Mike DeLuca, Scott Rudin, and Aaron Sorkin, who have all added to this project in numerous ways. Thanks also to Niel Robertson and Oliver Roup for much-needed guidance into the world of Silicon Valley. And many thanks to Barry Rosenberg, clearly the best at what he does.

Furthermore, this book could not have been written

without the generous, if sometimes reluctant, help of my numerous inside sources; though these sources have asked to remain anonymous, I have done my best to honor their cooperation by telling this story as honestly and respectfully as possible. I am an enormous fan of all of the characters in this book; I am in awe of their genius, and I am grateful to have been able to get a glimpse into a world of creation I'd never known before.

As always, I am indebted to my wonderful parents, to my brothers and their families. And to Tonya and Bugsy—I couldn't do any of this without you.

SECONDARY SOURCES

Baloun, Karel M. *Inside Facebook.* Victoria, BC, Canada: Trafford Publishing, 2007.

Brickman, S. F. "Face Off," *Harvard Crimson,* November 6, 2003.

Dremann, Sue. "In Your Face," *Palo Alto Weekly,* April 7, 2005.

Feeney, Kevin J. "Business, Casual," *Harvard Crimson,* February 24, 2005.

FM Staff. "How They Got Here," *Harvard Crimson,* February 24, 2005.

Forbes.com, "Facing the Future," September 13, 2006.

Greenspan, Aaron. *Authoritas.* Palo Alto, CA: Think Press, 2008.

Grynbaum, Michael M. "Mark E. Zuckerberg 06: The Whiz Behind thefacebook.com," *Harvard Crimson,* June 10, 2004.

———. "Online Facebook Solicits New Ads," *Harvard Crimson,* May 7, 2004.

Hoffman, Claire. "The Battle for Facebook," *Rolling Stone,* June 26, 2008.

Kaplan, Katharine A. "Facemash Creator Survives Ad Board," *Harvard Crimson,* November 19, 2003.

Lacy, Sarah. *Once You're Lucky, Twice You're Good.* New York, NY: Gotham Books, 2008.

McGinn, Timothy. "Online Facebooks Duel over Tangled Web of Authorship," *Harvard Crimson,* May 28, 2004.

McGirt, Ellen. "Hacker. Dropout. CEO," *Fast Company,* May 2007.

Milov, Sarah E. F. "Sociology of thefacebook.com," *Harvard Crimson,* March 18, 2004.

Neyfakh, Leon. "Columbia Rebukes thefacebook.com," *Harvard Crimson,* March 9, 2004.

O'Brian, Luke. "Poking Facebook," *02138 Magazine.*

Schatz, Amy. "BO, UR So GR8," Wall Street Journal Online, May 26, 2007.

Schwartz, Barry M. "Hot or Not? Website Briefly Judges Looks," *Harvard Crimson,* November 4, 2003.

Seward, Zachary. "Dropout Gates Drops in to Talk," *Harvard Crimson.* February 27, 2004.

Skalkos, Anastasios G. "New Online Facebook Launched," *Harvard Crimson,* March 19, 2004.

Tabak, Alan J. "Harvard Bonds on Facebook Website," *Harvard Crimson,* February 18, 2004.

———. "Hundreds Register for New Facebook Website," *Harvard Crimson,* February 9, 2004.

Vara, Vauhini. "Facebook CEO Seeks Help as Site Grows Up," Wall Street Journal Online, March 5, 2008.